Changing Identifications and Alliances in North-East Africa

Changing Identifications and Alliances in North-East Africa

Volume I: Ethiopia and Kenya

Edited by Günther Schlee
and Elizabeth E. Watson

berghahn
NEW YORK · OXFORD
www.berghahnbooks.com

First published in 2009 by
Berghahn Books
www.berghahnbooks.com

©2009, 2014 Günther Schlee and Elizabeth E. Watson
First paperback edition published in 2014

Library of Congress Cataloging-in-Publication Data
Changing identifications and alliances in North-East Africa / edited by Günther
Schlee and Elizabeth E. Watson.
   p. cm. -- (Integration and conflict studies : v. 2)
  Includes bibliographical references and index.
  ISBN 978-1-84545-603-0 (hardback) ISBN 978-1-84545-957-4 (institutional
ebook) ISBN 978-1-78238-329-1 (paperback) ISBN 978-1-78238-330-7 (retail
ebook)
  1. Group identity--Africa, Northeast. 2. Ethnicity--Africa, Northeast. 3. Africa,
Northeast--Ethnic relations. 4. Africa, Northeast--Social conditions. I. Schlee,
Günther. II. Watson, Elizabeth E., 1968-
  DT367.42.C53 2009
  305.800963--dc22

2009032993

British Library Cataloguing in Publication Data
A catalogue record for this book is available from the British Library

ISBN: 978-1-78238-329-1 paperback
ISBN: 978-1-78238-330-7 retail ebook

Contents

List of Illustrations

List of Maps,
Figures and Tables

Maps

Figures

Tables

List of Abbreviations

AK-47	Kalashnikov Automatic Weapon
AMDF	Arba Minch Development Farm
ASO	Anywaa Survival Organization
CPA	Comprehensive Peace Agreement
DC	District Commissioner
DFG	Deutsche Forschungsgemeinschaft
DSC	District Security Committee
EECMY	Ethiopian Evangelical Church Mekana Yesus
EPLF	Eritrean People's Liberation Front
EPRDF	Ethiopian People's Revolutionary Democratic Front
EPRP	Ethiopian People's Revolutionary Party
GOSS	Government of South Sudan
GPDUP	Gambela People's Democratic Unity Party
GPLF	Gambela People's Liberation Front
GPLM	Gambela People's Liberation Movement
HIPC	Highly Indebted Poor Country
IRC	International Rescue Committee
JEM	Justice and Equality Movement
KPDO	Konso People's Democratic Organization
KY	Kabaka Yekka
LRA	Lord's Resistance Army
LWF/DWS	Lutheran World Federation/Department for World Service
MEISON	All-Ethiopia Socialist Movement
NCA	Norwegian Church Aid
NFD	Northern Frontier District
NGO	Non-governmental Organization
NIF	National Islamic Front
NRA	National Resistance Army
NSCC	New Sudan Council of Churches
OAGs	Other Armed Groups
OLF	Oromo Liberation Front
OLS	Operation Lifeline Sudan
OPDO	Oromo People's Democratic Organization
PCOS	Presbyterian Church of Sudan
PRS	Proto-Rendille-Somali
RRA	Rahanweyn Resistance Army
SAD	Sudan Archive, Durham University
SAF	Sudan Armed Forces
SCC	Council of Churches in Sudan
SIM	Sudan Interior Mission

SNNPR	Southern Nations, Nationalities, and People's Region
SPCM	Swedish Philadelphia Church Mission
SPDF	Sudan People's Defence Force
SPLA	Sudan People's Liberation Army
SPLM	Sudanese People's Liberation Movement
SSDF	South Sudan Defence Force
SSIM	Southern Sudan Independence Army
SSLM	South Sudan Liberation Movement
SSUA	South Sudan United Army
TPLF	Tigrean People's Liberation Front
TTI	Teacher Training Institute
UDSF	United Democratic Salvation Front
UNHCR	United Nations High Commissioner for Refugees
UNLA	Ugandan National Liberation Army
UPC	Ugandan People's Congress
UPDF	Uganda People's Defence Force
WFP	World Food Programme
WV	World Vision

Introduction

Günther Schlee

One of the most basic questions of social science, namely who belongs to whom and why, continues to remain without an answer or to be answered in too many alternative ways, which amounts to the same thing. To explain collective identifications by group interests, as is sometimes attempted, falls short of a solution since, as groups emerge, their changing composition may lead to changing perceptions of shared interests. Any attempt to manipulate identities according to perceptions of political or economic interest has to start from pre-existing identifications. The options of those who have an influence on the form and change of social identities at any moment in time are limited by social givens. Identity politics is the interface of action and structure; it is where fluidity meets rigidity and either modifies it or breaks over it.

The substance of identity politics includes the possible ways in which people can claim to be the same as other people or to be different from them. These ways are myriad. Scope is one variant: identifications can be wider or narrower. Conditions can be identified under which it is advantageous for individual or collective actors to define wider identities which they share with others, for example to strengthen their own group or to widen their alliances when they feel insecure. In other conditions it may be preferable for a group to keep their own numbers small, when they do not wish to share their resources, or when they feel strong enough to prevail against their neighbours alone in a conflict and do not want any allies who would claim a part of the loot. Successful identity politics requires means of inclusion and means of exclusion, and the capacity to switch from one of these discourses to the other.

One way to move up and down in scope is to include wider or smaller units of the same kind, for example, units defined by criteria belonging to the same category, say linguistic criteria. Dialect differences, language and language family all provide linguistic criteria of identification. Within this category the options range from stressing minimal differences in dialect to underlining similarities between widely dispersed languages that are defined by linguists as belonging to the same families, and from this to postulate relatedness between their speakers (pan-Slavism, Turanianism, Bantu Philosophy …).

Anthropologists group other forms of belonging under the heading of descent. Descent reckoning can be of different kinds (different forms of linearity, non-unilinear systems …) or of different inclusiveness within one particular kind. Reference to real or putative remote ancestresses and ancestors normally implies the inclusion of more people within one's own group than those operating with shallow genealogies. Genealogical depth therefore often correlates with demographic strength.

Religion (or to use a less culture-bound term, possibly of easier universal application: belief system) is another such category. Religion, too, follows the segmentary principle. The primary identification can either be with a worldwide religious community or with a small sect or order. Similar considerations can be applied to all other subsystems of culture or complexes of symbols: they all provide materials that can be constructed as identity markers at various levels of inclusivity. Biological givens also come into it, most prominent among them pigmentation. Cultural definitions of skin colour categories vary widely. People who are perceived as white in one context are black in another and vice versa. In Sudan, a rather elaborate system comprising intermediate categories like yellow and red is at work. There are many ways to shift the line between 'us' and 'them'.

Taking narrower or broader criteria of the same general category to alter the inclusiveness of the intended collective identification is one way to formulate identities of different scope. Another way to do this is by changing the category of criteria. If one wants to enlarge one's group definition in a conceptual space in which religious commonality is perceived as more widespread than shared linguistic features, one might change from a linguistic group definition to a religious one. People who want to stress a more particular identity might move the other way in such a setting. That is, one switches from one category of criteria to another.[1]

Yet another way to move up and down in scope is by connecting categorical distinctions by different operators: 'and' or 'or'. To illustrate: there are more hot cakes and sweet cakes than cakes that are both hot and sweet. In the same way 'white', English-speaking Protestants who only accept other weakly pigmented persons of the same language and creed as being of their own kind apply a narrower type of identification than those for whom the presence of one or the other of these features already evokes a feeling of commonality.

Identities defined by different criteria typically do not replace each other but tend to coexist, often in some sort of hierarchy, albeit a changing and contested one. From a typology of forms of identification we can therefore move to a typology of forms of coexistence of identities. If one visualizes the categories of criteria by the dimensions of a graph, say by taking religious identifications as values along the y axis and ethnic identifications as values along the x axis, one will find some fields defined by given x and y values more populated than others. There are Christian Arabs and Muslim Arabs, but Buddhism rarely combines with an Arab identity, and so on.

Some ideologies postulate complete inclusion of identities defined by criteria belonging to one dimension into identities defined by criteria belonging to another dimension. Religious homogeneity can be postulated for ethnic or linguistic groups. Many Poles think that to be a proper Pole one needs to be Catholic.[2] More typically, however, identities defined by criteria belonging to different dimensions cross-cut each other, as we can see from the Christian and Muslim Arabs, Arab and non-Arab

---

1. This is one of the ways in which Elwert (2002) uses the term switching. What I call 'categories of criteria' or 'kinds of identity' here is the same as what I have called 'dimensions' of identification, on which particular identities are placed as 'values', in other contexts. No different meanings are implied. Different words are used to increase the chance of getting across ideas that remain the same.
2. Hann (1996, 1998).

Christians, etc. Even identities that are depicted by their bearers as belonging to the same general category can be found to cross-cut each other.

Ethnic and clan identity both tend to be stated in a descent idiom. Nevertheless, the same clans have been found in different ethnic groups in a number of cases. Cross-cutting identities have been primarily seen as binding forces, as cross-cutting ties.[3] More recently, they have been shown to be used in identity games of all sorts. In conflicts they have been found to have de-escalating effects in some cases and escalating ones in others, depending on factors that require further exploration.

The interior area of the Horn of Africa provides an ideal setting for studying these processes of identification and changing alliances and the way they transform over time. The southern and western parts of Ethiopia, the north of Kenya and Uganda and the eastern Sudan comprise numerous language communities belonging to three of the four macro-families of Africa (Niger-Kordofanian, Afro-Asiatic, Nilo-Saharan); they comprise Muslims, Christians and followers of a great number of other belief systems which, for statistical purposes, are often lumped together as 'animists'; they comprise political and military units of the most different types, ranging from 'nation' states to segmentary lineages and groups defined by sharing an age-grade organization. Some regions of Africa, like the area around Lake Chad, show a similar diversity, but it would be hard to find a region with a greater one.

This book brings together researchers who have worked in areas of study neighbouring each other in a relatively small part of the African continent. The aim is for the different cases, in addition to offering comparisons with each other, also directly to inform each other. The reasons for the regional approach are worth examining in more detail.

First, the ethnographies of neighbouring groups mutually complement each other. Views about neighbouring groups recorded by one ethnographer can be seen in the light of the self-image of these neighbours and their inverse views of the holders of views about them. Inter-ethnic interaction can be analysed in terms of the rationalities of both or all sides involved. Some of the groups studied by different researchers are at war with each other or share enemies. They might be exposed to the same policies or oppressive measures by the same states. By concentrating on a limited area we aim at exploring a mosaic of interrelated cases.

Secondly, the cases under study share much background information. While unrelated cases from different parts of the globe require separate introductions to their geographical and historical settings, much of this information is shared and does not need to be repeated if the cases under study stem from the same region. To proceed region by region rather than by hopping from continent to continent facilitates discussion of the thematic and interesting aspects much more quickly. In the context of the present volume, it is possible to outline once the ecological implications of living in the highlands or the lowlands of the Horn of Africa, the general history of state formation in this part of the continent, and some of the relevant state policies, although these issues are relevant to a number of the cases under study.

---

3. Gluckman (1966).

Thirdly, much of what has been called shared background information in the preceding paragraph, plus some elements of the thematic domain (for example, those topical elements that are found to be the same, or similar, in different cases under study), can, in a more formalistic, hypothesis-driven language, be referred to as constant elements. A situation characterized by a high proportion of constant elements can be called one of limited variation. In social systems that tend to be too complex for immediate recognition of which elements cause variation in which other elements, this limited variation is helpful for discerning covariation of potentially causally related factors if all factors do not vary at once. In experimental settings things are arranged in such a way that certain factors remain constant in different runs of the test. Here one speaks of controlled comparison. In the real world one has to wait to come across such cases of limited variation in the literature or in one's own experience, and can only hope to be fortunate enough to recognize them as such if they do arise. A regional focus increases the probability of such encounters.

Fourthly, in studying inter-group relations, the focus on a dyad, i.e. how A reacts to B and how B reacts to A, is useful. It is much better than an approach that concentrates on one group alone. The dyadic perspective, however, should not blind us to the fact that apart from A and B there are also C, D and E and that they may be affected by the relationship between A and B. C may profit from the fact that A and B weaken each other or A may gang up with D against B (see, for example, Schlee's chapter on the Garre, Gabra and Boran in this volume). The study of a plurality of neighbouring cases allows us to see regional patterns and interrelations and to discern indirect effects of interactions on third and fourth parties.

A regional focus alone, of course, does not lead to mutually complementary relationships between the single case studies, nor does it automatically lead to comparability. There are many other factors at play which mean that the ideal of such a regional project cannot be absolutely fulfilled. Ethnography is far from an exact science; the ethnographer brings to his or her study biases that make the research writings far from a simple, objective representation of those people under study. The structures, processes and perspectives of a particular group (for example, group A above) are also complex and varying, and it is difficult – or impossible – to write a definitive account of group A's perspective on group B. Each of the researchers contributing to this volume has spent time working in the region with his or her own research focus; it is not necessarily possible to draw direct comparisons between groups from these diverse research endeavours. Thus the chapters of this volume do not come together like a patchwork quilt, with all the edges of each story lining up uniformly and securely. There are gaps and disjunctures and some stories point in different directions. But this is not necessarily a bad thing: each study does shed light on each other, and on the processes through which identity is formed, changes and endures, and its strength in shaping political outcomes, livelihood opportunities and conflicts. These issues are, in the present day, extremely pressing, and the diversity of chapters included in these volumes allows different perspectives on these issues to emerge. The volumes include chapters from scholars from Japan, Ethiopia, Sudan, Europe (UK, Germany, France, the Netherlands) and the USA. (Their nationalities may not be the most important identifications among them: common forms of identity based on place or work (e.g. Kenya, Uganda, Ethiopia or Sudan), or with pastoralist or agricultural groups, or theoretical or disciplinary perspective, or age or gender may be a better way of describing the different groups.) The contributors also include scholars who have many years of experience in the 'field' to reflect on and new researchers who are just finishing their first substantial piece of research. The diversity of contributors helps to enrich this account of the region and its identity processes and politics.

In order to facilitate discussion amidst this diversity, this study needed not only a shared interest in a region, but also some agreement about what kind of thing to look for and a shared terminology for talking about these things. In other words, we needed a basic set of questions and theoretical framework. An outline that roughly corresponded to the first paragraphs of this introduction was sent along with the invitation. Contributors were asked to focus on changing systems of identification, on inter-ethnic relations, including the relation to the state or states, the ethnic make-up of states or the equations drawn between 'ethnicity/ethnicities' and 'nation/nations'. In order to provoke the emergence of patterns, we had to make them address comparable or mutually complementary issues.

As our ambition was to generate advances in general anthropological theory in addition to adding facets to a regional ethnography, Liz Watson and I decided to make the resulting volumes readable also to those who are not area specialists. We therefore wrote a second introductory chapter, to follow the present one, which is more theoretically oriented, on the geographical and historical setting of the case studies collected in these volumes and on political developments at the national and international levels in the different countries where the research areas are located. In the present volume, the setting is Ethiopia and Kenya; Volume II refers largely to Sudan and Uganda, and a similar general introduction to those contexts is included in that volume. In this volume, the general historical and geographical introduction that follows the present chapter contains much that is already known to the area specialist, who might therefore skip it. The non-area specialist might do the opposite. If he or she finds that the remainder of the present chapter contains too many names of peoples, language families or cryptic references to historical events, s/he may read the historical and geographical introduction ('Space and Time') first and then come back to the present chapter.

To have more than one editorial chapter also gives us the opportunity to summarize some of the literature. Not everyone who has done relevant research on our topic has been able to contribute to this volume. The editorial chapters were therefore also planned to provide space for a fresh look at some of the relevant older literature. The review of themes in identity politics may also provide a useful theoretical introduction for the non-specialist.

Cross-cutting Themes in Identity Politics

In the remainder of this chapter, I look at four cross-cutting themes that are central to the topic of identity politics in North-East Africa, and which are discussed in more detail in the chapters that follow. The first is the theme of group identity, dominated by notions of 'us' and 'them'; the second explores the role and perception of conflict; the third revisits debates about generation-grade and age-grade systems, prominent in the Horn of Africa; the fourth examines links between identity, inter-group relations and population and natural resources. These summaries are designed to provide some of the common background to several of the studies.

Similarities and difference

Many of the chapters in this volume explore how ideas of similarity and difference between groups change over time, and the way in which they are negotiated by local groups and individuals to produce identities. A number of contributors (Amborn, Abbute) follow Elwert (1989) in using the terms 'we'-group and 'they'-group, which may also be inspired by social psychology. These terms evoke an emotional load of social identities with solidarity in the case of 'we'-groups and distance in the case of 'they'-groups. As such emotional charging occurs, and to some extent it often does, this terminological use is perfectly legitimate. The emotions involved and their essentialist ring should not make us forget that the use of such pronouns is always situational ('we', the parents, versus 'them', the teachers; 'we', the Germans, against 'them', the French; 'we', the Protestants, against 'them', the Catholics, etc.). Moreover, even if referring to the same collectivities, their emotional load may go up and down over time ('we', the supporters of a given football team, may have a good feeling for one afternoon). Where emotions are not the topic, we prefer to speak of identity and difference, and the categories and groups defined by them, without the adjectival use of such pronouns.

The strength of ideas of similarity and difference to others changes over time. Identities are constructed through the interface between individual and group action and changing structures. These structures, as outlined already, include indigenous forms of descent or belief, and wider structures such as religions, government institutions and policies, economic systems and environmental context. New developments such as the influx of guns or other ideas and technologies may also have an impact on notions of sense of self and others and on alliances. As structures and forms of individual and collective action interrelate, stronger and weaker forms of identity may emerge. It is possible that, as stronger – more emphasized – forms of identity are produced, incidences of conflict are more likely to occur. The process may also take place in reverse, however: incidences of conflict may make certain identities stronger and their boundaries more clearly defined and relevant to everyday lives. The history of conflict in an area, and the way in which it is perceived, also influences outcomes, and it is to this I now turn.

Emic ideas about warfare

In the area under study, wars are often perceived as periodically recurrent and quasi-natural and inevitable. In Europe such ideas have also persisted until recently. In my family things were rarely thrown away. I remember reading as a child, around 1960, an article about dogs in a children's periodical from the 1870s or 1880s. These dogs were trained to find wounded people and would be very useful 'in the next war'. The next war was something to be expected quite naturally after a period of peace. After rain comes sunshine, and after sunshine comes rain, and, in a similar way, war and peace were thought of as alternating. Born after the disaster of the Second World War and raised under the threat of nuclear overkill, my generation might be the first generation of Germans that has been brought up to think of war as a completely unrealistic option and of preventing a war rather than fighting wars being the purpose of the military.

Let me now adduce some evidence from North-East Africa, first for war being thought of as a natural phenomenon and then for war being thought of as periodically recurrent. In Lokiliri, the main village of the Nilotic-speaking Lulubo of southern Sudan, the most important ritual function is held by the King, the Master of the Rain. He is responsible for rain in the whole of Lokiria. For parts of the area, there are also minor Rainmakers from specific clans.

Below the King/Rainmaker(s) there are holders of other specific powers. They are known as the 'Masters', for example, the Master of the Land, the Master of the Mountain, the Master of Locusts, the Master of Tsetse Flies, the Master of Leopards, the Master of Lions, the Master of Grain, the Master of Worms, the Master of Winds, the Master of Birds.

In some of these categories there is more than one office holder. One Master of the Land from the Loo clan controls a large area; another from the Mondugi clan and yet another from the Onyoko clan control other areas. 'Control' means that they have to 'cook the land', i.e. to prepare it ritually for cultivation. In other categories there is just one Master: the one from the Ongairo clan is the only one who can threaten the community with a plague of locusts; the one from Okare is the only one who has power over tsetse flies, and thereby has general responsibility for the health of cattle (Simonse 1992: 264–66): 'Each of these Masters is believed to control a specific threat to the well-being and survival of the community ... Most of the offices concern powers over "natural disasters" (if we include disease in the realm of nature), except for the power over spears and arrows' (Simonse 1992: 266). The powers over spears and arrows are held by the blacksmith clans Okare and Kutunot, and comprise both the power of deflecting enemy weapons and that of 'sharpening' the weapons of Lokiria men and making them hit their targets: 'It is significant that enemies and natural disasters are put in the same category of phenomena' (Simonse 1992: 266).

The Cushitic-speaking Rendille of northern Kenya have nine clans, which are divided into different sub-clans. Some of these, those who are *iibire*, have a potent blessing and curse, while others can only trust in God and are therefore called *Waakh kamur* – 'God is rich/mighty'. The *iibire* clans or sub-clans (there are clans with both *iibire* and *Waakh kamur* sub-clans) have different vehicles for their curses. If someone has a skin disease called *nabhar* (probably a fungus), he is believed to have been cursed by Dubsahai; snakebite is attributed to the curse of Rengummo. Saale has a special relationship to the rhinoceros, Tubcha to the elephant. Somebody cursed by a member of these two clans may be trampled by the respective animal. There is one group of *iibire*, the sub-clan Gaalorra of Gaaldeylan, whose curse vehicle is the horse. As the horse is a domestic animal, and in most cases harmless, it appears strange to find it in the same category as the dangers of the wild. But the horse here stands for Boran cavalry. For the Boran Oromo the horse was an effective means of raiding: galloping through a herd of camels would make the camels stampede and run with the horses. As the 'horse' stands for 'enemies', here again we find enemies put in the same category as natural phenomena.

Members of Gaalorra would pray for the Rendille if under threat of war, and they would curse the enemy and lead them astray. But they can also use their curse vehicle, the 'horse', against other Rendille in the same way as Rengummo can send a snake or Saale can send a rhino against those who have incurred their wrath. Somebody cursed by Gaalorra is believed to be exposed to enemy raids.

A comparable situation exists between the Lulubo and the Rendille, where powers are passed between descent groups. These are clan-specific powers. A difference seems to be that in the Lulubo case a given power belongs to one ritual office held by one man at a time, while among the Rendille all members of a clan hold the occult power of that clan in various degrees. Some people are believed to have a more efficient blessing and curse than their clan brothers, and this is believed to depend on their ritual purity (keeping clan-specific food avoidances) and the will of God. God is thought to be able to accept or reject prayers as he likes, including curses, which are seen as prayers for harmful effects: he is not seen as being accountable for the way in which he bestows his favours (Schlee 1979, 1994a [1989]).

These examples, which could be multiplied by evidence from other groups, show that enemies and war belong to the realm of the normal, that they are facts of life like locusts, skin diseases or poisonous snakes. The second point I wanted to substantiate is that war is often thought of as periodically recurrent.

Among Lowland Eastern Cushites generation-set systems of the type called by the Oromo term *gada* are extremely common. *Gada* systems differ from other generation-set systems, like those common among Nilotes, in that they are based on an elaborate calendar and a numerical order. Sons, at least the elder sons, who are meant to fit into the ideal unrolling of the system, are initiated a given number of years after their fathers. Rendille youths are circumcised and thereby recruited as a warrior age-set forty-two years after their fathers. Initiations take place once in fourteen years, so that a generation consists of three age-sets, 1, 2, 3, followed by the next generation, where the sons of 1 are succeeded by those of 2, and the latter by the sons of 3. Historical events like wars and famines are believed to recur in a double generation step, i.e. grandsons share the fate of their grandfathers.

The Rendille say that the age-set Ilkichili, who were warriors from 1965 to 1979, is an 'age-set of blood'. In fact the number of wearers of killers' insignia (a special arrangement of beads as a necklace and brass bracelets) among them was much higher than in preceeding or subsequent age-sets. The killer status of these distinguished warriors was achieved through mutual raiding with Gabra and Turkana, and the number of Rendille victims probably roughly corresponded to those of the enemies killed by Rendille. Ilkichili are believed to have repeated what occurred also to their grandfathers, Dismaala, warriors from 1881 to 1895. In that period misfortune struck the Rendille to such an extent that the name Dismaala was struck from the list of names that periodically recur as names of Rendille age-sets (Schlee 1979).

In *gada* systems the names of age-sets that cyclically recur do not necessarily correspond to the number of age-sets per generation or double generation, so that a given name might be allotted to a given line of age-sets related by descent (fathers and sons) only after a major cycle (number of names × number of age-sets per generation = number of age-sets after which a given name comes back to a given descent line). Among the Boran the coexistence of such major and minor cycles has given rise to the specialist function of *ayaantu*. An *ayaantu* is a diviner who makes predictions based on the cyclical reoccurrence of past events. For each war and each raid the Boran are able to tell which war of the past has 'come around'.

Of course, also for the peoples under study, regarding war as natural and cyclically recurrent is only one of several ways of looking at it. This fatalistic perspective might help people to accept the hardship and losses associated with war. But they are not fatalists in all contexts: they evade danger, realign their alliances and complement war with politics. It is with this active dealing with both hostile and peaceful inter-ethnic relationships and with the calculations behind it that these volumes are primarily concerned.

Age grades, lineages and 'predatory expansion'

Divination and the interpretation of history as recurrent wars and disasters by no means provide the only link between generation-set systems and warfare. Tornay (in this collection) describes the generational system of the Nilotic Nyangatom, Turkana and Toposa as an instrument of 'predatory expansion', taking up a famous phrase used by Sahlins (1961) with reference to the segmentary lineage system of the Nuer. The two forms of organization are normally seen as complementary to each other: one's segmentary position is determined by references across the generations, diachronically, back in time to one's lineal ancestors, while generational and other age-grading systems proceed synchronically and cut across descent lines to group people together who have the same value on the timescale: in terms of either generation, part generation or actual years. But, Tornay suggests, these two complementary forms of organization might in other ways be equivalent, namely in their recruiting potential for violence.

Without getting entangled in earlier debates about the segmentary lineage system,[4] it can be said that the Nuer, indeed, have used their lineage system successfully to build up demographic and military pressure. The Anywaa describe the Nuer in precisely Sahlins' terms. Some feedback of scholarly debates into local discourse might have taken place here. There are societies that largely conform to Evans-Pritchard's (1940) model of segmentary lineage systems, in spite of all efforts to discredit it; some are even more 'predatory' than the paradigmatic case, the Nuer themselves. Not all of them are engaged in expansion all the time, but it can be shown also in cases other than the Nuer that a segmentary lineage organization combines well with an expansive dynamic. In Volume II of this collection of essays, both Dereje and Falge deal with the Nuer.[5]

---

4. For fuller references see Schlee (2002).
5. Schlee (1988) has compared the Somali, who have a quickly segmenting lineage system, with the Rendille, who have clans which they believe to be immutable and bestowed with different personalities and ritual powers. The former pursue maximization strategies both in their livestock management and in that of humans, while the latter show conservative and self-regulatory tendencies in both fields. The Rendille adjust to limited resources. The Somali use them up and then conquer new ones.

Nuer and Somali lineages can be joined by strangers; there are forms of individual adoption and the association of entire groups. The generation-sets of the Nyangatom do the same. Tornay himself was accepted into the Elephant generation-set back in the 1970s. He did not have to struggle to achieve this. He bought a bracelet of the type worn by the Elephants, without knowing its implication, put it on, and that was it: he was dragged by a bystander into the Elephant dance and has been an Elephant ever since (Tornay 2001: 12f.). The incorporation of strangers to swell one's ranks is an indicator that indeed a growth-oriented and expansive rather than a discriminatory dynamic is at work here. Both segmentary lineage systems and generation-set systems have the potential to produce mass effects, to integrate and mobilize many people.

Generation-set systems and the activities associated with them are by no means purely a military organization but pervade all spheres of life. The *gada* system of the Boran has been described by Baxter (1978) as having mainly ritual, fertility-oriented functions rather than military ones. But this prevalence of peaceful attributes might also be a consequence of the Pax Britannica that prevailed during Baxter's field research. There is evidence that the Oromo expansion since the sixteenth century was made possible by a reform of their *gada* system to make it fit with the actual age of the people who joined it, so that those recruited into the warrior age-set were actually young men. Rigid systems have a tendency to get out of step with the demographic reality and to require reform or ad hoc adjustments from time to time (Asmarom Legesse 1973). Ritual warfare and the acquisition of booty, including the genitals of slain male enemies, have been a major factor in the ethnogenesis of the present groups found in northern Kenya, some of whom derive from people who tried to withdraw from this pressure, others from those who accepted Boran hegemony and paid ritual tributes to them (Schlee 1994a).

The juxtaposition of fertility-oriented rituals and warfare in the last paragraph should not suggest that the two have nothing to do with each other. On the contrary: not only among the Oromo[6] but also among many of their neighbours, the killing of enemies and the acquisition of genital trophies are important for guaranteeing the fertility of the land, the cows and women. The Hor of south-west Ethiopia (often referred to as Arbore) carefully distinguish among the ethnic groups neighbouring them the enemies who are good to kill and those who are not. The former are more respected as enemies and better for the fertility of the land (Tadesse Wolde Gossa 1999).

Generation-set systems form regional clusters. The *gada* systems of the Eastern Cushites have already been described as sharing the characteristic of following a strict numerical[7] order. They also differ widely in the length of cycles of initiation and in the number of sets per generation and other fundamental structural features. There

---

6. For a recent summary of the ideas surrounding ritual killings and fertility, which even pervade the modern development discourse in Oromo, see Zitelmann (1999). Also our book on *Islam and Ethnicity in Northern Kenya and Southern Ethiopia* (Schlee and Shongolo, in preparation) will contain a section on 'The killer complex'.

7. In emic perception it is often viewed more as a sequence of named units of time.

are, however, forms of ritual cooperation between them: the Gabra, with their different systems, can only start their ritual cycles after the senior Gaar phratry has done so, and these need the gift of a heifer from the Boran to trigger their system into motion. In the past, the Garre had to give ritual implements like a certain type of cloth to the Boran. To some extent the ritual well-being of the 'hegemonic' Boran depended on Garre participation in their rituals (Schlee 1998b). My own chapter in the present collection shows how far the Garre/Boran relationship has declined since those days of ritual cooperation.

A feature that many age-grading systems of this area share across the Nilotic/Cushitic divide is not only that they comprise offices that have the positive attributes of leadership and convey honour, but also that they have what might be called 'negative offices'. The Rendille have the *dablakabiire* and the *arablagate*, the Nyangatom, Karimojong, Koegu, Kara and Pokot the *asapan* (Kurimoto and Simonse 1998). In all of these cases, certain people are sought out and then the childish attributes of an entire age-set to be promoted to a senior grade are unloaded onto them. In this way misfortune that might otherwise befall the age-mates is directed towards the appointed individual. Ideally the man in question or his family should be richly compensated, but in practice this is often neglected[8] and the harmful rituals are performed with undisguised force. This institution may therefore lead to intra-societal conflict or to fission.[9] One may see similarities between these generation-set systems and the offices they provide, both in terms of leadership and for attracting misfortune (the lightning conductor function), and the scapegoat kings described by Simonse (1992) among the Bari, Lulubo, Lokoya and Lotuho further to the west, towards the White Nile. Here the ambivalence resides in the same person. These kings combined ritual with political power (before the British introduced the distinction between Rainmaker and Chief) and both forms of power brought with them the risk of violent death, because any misfortune was attributed to them. 'Divine kingship' is the catchword under which this complex of ideas has been studied on a worldwide scale.

Another feature shared by many Nilotic and Cushitic age-grading systems is that they do not provide the elders with effective control over warriors. This is not only so when the younger, fighting males belong to modern organizations with their own hierarchies (see the disagreements between *gada* authorities and the Oromo Liberation Front [Shongolo 1996]) but also in spheres entirely untouched by modern politics. The British colonial administration in Kenya bitterly complained

---

8. The *dablakabiire* of the Rendille should be given a new sheet and be allowed to marry ahead of his age-set and the girl of his choice, as a compensation for henceforth being regarded as silly. The last time this happened, the man was abandoned after the ritual had been performed, and he resorted to murder and ultimately suicide (Schlee 1979). Among the Nyangatom in 1980 there were complaints that the *asapan* of the generation-set of the Ostriches had not been chosen properly. It was usual to buy a man with substantial livestock payments for this purpose, but this one had been taken by force and he subsequently turned mad and died in the bush (Tornay 2001: 314).

9. See the chapter on 'The Conflict about the Marriage Rituals of the Age-set Ilkilchili' in Schlee (1979).

about the truculence of the Samburu, who never kept their peace agreements. When the elders had concluded peace and sanctified it by solemn prayers, the warriors continued with their raids (Spencer 1973). By the time the British left Kenya they might have found out that the people who make peace among the Samburu are not the same as those who make war, and that there is no chain of command between the two. Rendille elders can conclude peace but they have little influence on when the war starts again. Recently, major clashes with huge losses on all sides started with the spontaneous decision of a Rendille warrior to cut the throat of a Boran herdboy with whom he was talking and to take the genital trophy and drive away his cattle.

Struggle and scramble

At the very beginning I said that I wanted to shift the emphasis away from resources, which have too long been the sole focus of explanations of conflicts, and to ask who is against whom, or who is with whom, which criteria of identification are used, and how the sides in a conflict are defined. All these can push the number of partisans in a conflict up or down. But one aspect of resources is important for just this numbers game and cannot be neglected by the incipient size theory of identification: this refers to the matter of not so much what the resources are but how people compete for them. In this competition, the nature and quantity of resources play a vital role. There are basically two types of resource competition: the struggle and the scramble. Let me illustrate the difference.

> Case no. 1
> We are two and there is just one bowl of food. We fight over it. I win and eat the food. This is a struggle.

> Case no. 2
> It is again you and me and the bowl. We each have a spoon and eat from the same bowl peacefully. I can eat faster and get more. This is a scramble.

The chapter about Mbororo in the present collection (Dereje and Schlee, Volume II) depicts a peaceful and evasive resource-use strategy of this group of pastoral nomads, who, after all, have managed to come all the way from West Africa without getting killed. They stay far away from farmed areas to avoid litigation involving their cows damaging someone's crop. They move into the grazing areas of other groups, including those reserved for later grazing, but by the time people complain they have moved on.

Their cows need these resources. They cannot stay around the villages for as long as those of the Arabs or Oromo. They always need lots of fresh grazing. They are a large, high-performance breed with a high energy demand, quite different from most African cattle, which survive on a low-energy diet. They are long-legged and good walkers. In other words, these cows are eating machines, designed to get the best fodder far and wide and to get it fast. They are the bovine equivalent of the Somali camels, which I have compared with the smaller and hardier Rendille camels along similar lines (Schlee 1988).

With their eat-and-run tactics the Mbororo have fared well for a long time. It often happened that people complained about the overgrazing of their cows and expelled them, but then they just moved on and might be back some years later. In our joint chapter Dereje describes how this relatively peaceful and evasive strategy of resource acquisition has broken down for those Mbororo who live among Nuer. Tightened government control has managed to keep them out of Ethiopian Gambela, and in the adjacent parts of Sudan they were, in a way, captured by a faction in the civil war, which has made them pay a heavy fee for their use of pastures without providing them with adequate protection. Now not only are they arms traders, which they might have been for a long time, but they also present themselves as heavily armed and ready to fight. They are cornered. They have been dragged from the scramble into the struggle.

If we compare scrambles with struggles, numbers play a role in quite different ways. In a struggle one needs to be strong enough (and that might mean numerous enough) to win, but one should not have too many friends and allies on one's own side when it comes to sharing the loot. In a scramble, it might be an advantage to be small from the start: to be inconspicuous, to eat fast and to get away.

In a macroscopic historical perspective, looking at larger groups over a long period of time, groups of people might be squeezed out from the use of certain resources by others who are faster and more efficient in the use of those resources. They might even be demographically reduced or dislocated in the process without a single drop of blood being spilled. For them, it does matter whether that other ethnic group with whom they are peacefully, if involuntarily, sharing their resources is large or small.

For the individual fast eater who engages in the scramble mode of resource competition, large numbers are invariably a negative factor. If we all eat from the same bowl and the food is limited, it matters little whether those eating with me (or one can say against me) are of my own kind or different; there are no friends or foes here, only competitors. Speed in paying one's compliments and getting to the next bowl might, however, be an advantage.

Where numbers are used in a scramble, we are not dealing with a true scramble but with a struggle in disguise. If the hosts keep quiet because the uninvited guests are too many, it is, in fact, the threat of latent violence they possess which does the job.

These four themes are prominent in theories of identity politics and are discussed in more detail in the chapters that follow. They are by no means the only theories and approaches discussed, nor are they mutually exclusive: combinations of these processes may be at work in different contexts. Identity politics are not becoming less powerful in recent years; if anything, in relation to global processes, they are becoming more evident and extreme. The chapters that follow in the two volumes aim to improve understandings of these processes, the factors that are contributing to them and the ways in which forms of identification, lives and livelihoods of people in this region interrelate. Although based on one geographical region, it is hoped that they may also provide insights of value for understanding processes of identification and alliance formation in other parts of the world.

Space and Time: Introduction to the Geography and Political History

Günther Schlee and Elizabeth E. Watson

The chapters in these volumes deal in different ways with the interactions between history, contemporary political manoeuvrings, geographical locations, the communication of knowledge and ideas, collective memories and power relations. Identities and alliances are produced through the interactions between these endogenous and exogenous structures and agency. Overall, these interactions can be summarized as being strongly influenced by the shifting and unstable relations between space and time. Historically in this region, these processes have been dominated by relations between centres and peripheries: the centres have promulgated policies and promoted 'developments' that have been accepted, resisted or transformed by even the smallest and seemingly most 'remote' group of people. In recent years, improved communications and decentralization programmes have challenged the dominance of centre-periphery relations: for example, Donham (2002: 2) argues for Ethiopia that 'a hierarchical arrangement of cores and perhipheries, apparent to all and inscribed upon geographical surfaces', has been replaced with 'a more open series of interactions drawing on partially shared and intersecting "ethnoscapes" of the imagination'. The friction that constrains movement through space has not been completely overcome by new technologies, however, and old and new hierarchies are in place. The ways in which the processes of identification and alliance construction are mediated by the changing relationship between space and time require exploration in their particular contexts.

This chapter introduces the settings of the chapters that follow and some of the themes they discuss. It also provides a basic outline of the political histories of the countries in which the chapters in this volume are set: Ethiopia and Kenya. A similar introduction to the political history of Sudan and Uganda, in which the chapters in Volume II are set, is given at the beginning of that volume.

Space: Geographical Settings

The location of the groups discussed in this volume and the next can be seen in Map I.1.

From a macroscopic geographical perspective, the area under study is on and around the watershed between the Atlantic and the Indian Ocean. The Nile flows into the Mediterranean Sea, a bay of the Atlantic, and the River Juba into the Indian Ocean. In more humid eras, the River Awash was part of the Indian Ocean system

Map I.1 The Horn of Africa and the approximate location of the people who feature in these volumes.

and the River Omo of the Atlantic system via the Nile. Now neither of them has an outlet into the sea: the Omo flows into Lake Turkana; the Awash dries up in Djibouti.

A look at the river basins is useful for understanding the human geography. Water flows are often parallel to the movements of people and trade goods. Traffic was along the rivers as far as they were navigable and further up, in the Ethiopian Highlands, along the plateau and ridges separating them. Moving at ninety degrees to these axes, across highlands and rivers, would have meant overcoming deep gorges, the most impressive that of the Blue Nile, traversing malarial and 'spirit-haunted' lowlands abhorred by the highlanders, and requiring a great deal of unnecessary mountaineering.

Until 1991, the provincial boundaries of Ethiopia took these 'givens' into account. They were designed to include people from sub-centres, and for the sub-centres to be reached from Addis Ababa at the centre. Some of these boundaries go back to the realms of the conquerors, to the generals of Emperor Menelik at the turn of the twentieth century. Their spheres of control followed the lines of traffic. In the 1990s, administrative boundaries were redrawn 'to conform to lines of supposed cultural difference' (Donham 2002: 6). This had the consequence that the Yem people of the former Kafa province now have to travel via Addis Ababa to reach Awasa, the capital of their new regional state, the Southern Nations, Nationalities and Peoples Regional State (Popp 2001). The New Beni-Shangul-Gumuz State on the Sudanese border straddles the Blue Nile. For officers from the capital, Assosa, to reach Metekel in the Gumuz area on the other side of the Nile (former Gojjam) it is necessary to go via the bridge at Neqemte: a trip of 1,250 km (Young 1999: 342). From Neqemte it is just as close to Addis as it is to Metekel. For this reason (and possibly other reasons) delegations from the two parts of Beni Shangul/Gumuz like to hold their joint meetings in Addis Ababa. Many of the officials have little aversion to absences from home or office and long-distance travel, because they are paid expenses on a per diem basis.[1]

The major division, the watershed between the Atlantic and the Indian Ocean, found its political equivalent in the first half of the twentieth century in the British and French spheres of economic interest. The French built the railway from Addis Ababa to their colony Djibouti. This railway follows the Awash Valley and directs the trade to the Indian Ocean and Red Sea, via the Gulf of Aden. The British were then in control of Sudan. They had an enclave in Gambela on Ethiopian soil and tried to direct as much trade as possible from western Ethiopia via the Baro, the Sobat and the White Nile to Khartoum. In the dry season, when the Baro was low, the steamer services were replaced by lorries on dirt roads parallel to the river. A much smaller proportion of the trade with Sudan went along the Blue Nile (Bahru Zewde 1976). Transport was mostly by road and, on the Sudanese side, by rail. Uninterrupted boat travel on the Blue Nile was not possible after the construction of huge dams for irrigation schemes, the largest of which was the Gezira scheme, in the 1920s.

Exports from western Ethiopia comprised coffee, gold and, off the controlled routes, well into the twentieth century, also slaves (Triulzi 1981). International companies struggled over platinum mining licences in the Jimma area: the 'white gold' (Cerulli 1933: 73–79). In Sudan the waterways were important, as were the roads across the desert. Even before motorization these included the caravan routes, which made short cuts across the bends of the Nile, and the famous *darb al arba'iin,* the 'forty days route' directly from Darfur to Upper Egypt. The lowlands provided no obstacles except for the vast distances.

The experiences, livelihoods and environments of each group discussed vary tremendously, across and between short distances. However, some patterns can be

---

1. Günther Schlee's notes about a conversation in Beni Shangul, Ethiopian Diary 2001/2, 24 November 2001, MPI for Social Anthropology, Halle/Saale. (http://www.eth.mpg.de/subsites/schlee_tagebuch/index.html)

identified in relation to particular geographical matters: the environment and livelihood practices of the people involved, and the degree of their remoteness or connectedness to the state and to wider processes. Certain themes regarding the social processes of identification and alliance building are also more prominent in certain areas than others, showing that geography relates to more than just soil, rain and access to markets.

The people discussed in these volumes practise a wide range of livelihoods, from very intense agriculture to extensive pastoralism and hunter-gathering (or a combination of these). Livelihood choices are not necessarily determined by environmental matters, such as the interrelated rainfall and altitude, but they may be constrained by them. Some people discussed in these volumes are living in towns or refugee camps, and depend for their living on very different kinds of resources: they depend on obtaining resources from the state, from private enterprise or from the international humanitarian community. The degree of mobility or fixedness of everyday livelihoods is shown here to make a difference to the sense of self, the way a person or group relates to space and place and to the wider world. The reasons for that relative mobility also make a profound difference to the nature of the experience and its impact: migration across a border in search of a refuge or constraint in a refugee camp generates a very different relationship to space and place from that of groups who have historically moved over large distances in search of pasture for animals. Some, but not all, of these groups are experiencing conflict and concomitant displacement.

The 'degree of connectedness' to wider processes also determines the geographical scope in which the processes of identification and alliance formation are played out. Here space is not fixed but elastic and relative. It is compressed when, for example, people in refugee camps on the Ethiopia-Sudan border are in contact with relatives and donors in the United States (see Falge, Volume II). Some of the groups discussed in these volumes have only slight contact with their national or regional governments, which do little in the way of providing services or intervening in local conflicts. Others, almost as far away as these from the centres as the crow flies, may be better connected in terms of infrastructure and their access to state resources, and their concept of themselves and their place in the world may vary accordingly. These chapters explore how modernity has come to (and been appropriated by) people in terms of religion and education, or in the form of the gun trade, all of which have reoriented people's senses of themselves and their geographies. Access to new weaponry is related to the position of many of these people near international borders where illicit trade takes place, and to broader national and international political transformations, particularly the end of the cold war. Access to weapons and the militarization of society that has accompanied it have led, as Hutchinson has written, to 'a radical reconfiguration of power relations' (Volume II; see also Hutchinson 2000). The proliferation in the number of guns has changed constructions of masculinity and led to a transformation in relations between men and women (Masuda, Abbink, Volume I). It has also challenged other traditional structures and hierarchies of power, such as those between old and young. This has reduced the power of conflict resolution mechanisms, making the multiple problems suffered by the people frequently more entrenched and protracted.

The case studies have been organized into two volumes. In the first volume, the majority of the cases are from Ethiopia, although two are from northern Kenya. The second volume includes the chapters relating to people who are situated in Sudan, or those who have migrated from (or to) Sudan, some of whom may be living in refugee camps in surrounding countries. An exception to this in Volume II is the chapter by Gray on the Karimojong of northern Uganda, which explores in depth the experiences of war in recent years, through the words and memories of Karimojong women.

In this first volume, the chapters have been divided broadly into four sets that correspond to area of location and also, roughly, to livelihood type. The first set compares the experiences of three groups of people from the Omo Valley in the far south-west of Ethiopia. This area is inhabited by several small groups of people, the populations of some of whom number only a few hundred or thousand. Linguistic differentiation is high. Some languages belong to the Nilotic family, of which Nilo-Saharan is the wider affiliation. Others are classified as Omotic, and yet others as Cushitic. Omotic and Cushitic are different branches of the Afro-asiatic macro-family. Despite the small sizes of some of the groups, they have retained strong senses of identity and distinctive ways of life. They are well known in the anthropological literature and have also featured prominently in the coffee-table photograph books of the area (see Abbink, Volume I). The livelihoods of the majority of the people in this region are based on agro-pastoralism: they keep (and raid) cattle and cultivate land. The Nyangatom (Tornay) carry out riparian agriculture on the land that surrounds the River Omo, using the flood waters; the Suri and the Banna (Abbink and Masuda, Volume I) augment their pastoralism with highly variable rain-fed agriculture. They combine these activities with a certain amount of opportunistic hunting and gathering.

The picture that emerges of the Lower Omo Valley in these chapters is of people who have very uneven relations with the state. On the one hand, they are remote from the state, and the state provides them with few resources. On the other hand, their proximity to national borders, which are at times heavily defended by national or rebel armies, and which the people cross from time to time, can only make them acutely aware of the state, even if it is not very empowering for them. Insecurity and displacement are prominent features of life here: there are raiding and counter-raiding between groups, and the victors gain access to territory, cattle, women and resources; the vanquished are forced to seek safer areas, frequently finding themselves displaced from their symbolic homelands (Abbink, Volume I). Although violence has always featured in life, access to modern weaponry has made matters worse, as 'fear of being raped or killed by our enemies' is now common, and 'an alternative mode of sudden warfare and fragile peace remains the normal mode of communication between the peoples' (Tornay, Volume I).

More than any other peoples presented in these books, the people of the Lower Omo have been viewed by northern Ethiopian administrators and policymakers, as well as foreign tourists, as 'really primitive tribes' (Abbink, Volume I). In the past, in their encounters with the state, they have been treated much more as subjects rather than citizens (Mamdani 1996; see also Tadesse Wolde Gossa, this Volume). Since the Ethiopian government policies of the post-1991 period, which have emphasized the rights of each nationality to self-determination, minority status combined with claims to being autochthonous have provided them with a certain amount of political capital (Watson, Volume I; Dereje Feyissa, Volume II). As Abbink (Volume I) explains, 'their peripheral position as an "ethnic minority" or "nationality" has now become a kind of privilege'. The extent to which the people of the Lower Omo Valley have been able to take advantage of this reversal in the power politics of state is explored in this first set of chapters. Where the state has failed to serve as the main means of communication between these people and the wider national and international community, or as provider of resources, Christian churches (foreign-based and indigenous) have played a powerful role, providing resources and sources of meaning for people who find themselves in an increasingly liminal situation (Tornay, Volume I; also Falge, Volume II). The government's ethnic federalization programme has brought certain new dynamics to the politics of ethnic identification and alliance; guns have brought the means through which those identities are performed and negotiated; and men and women, young and old, have engaged differently with the possibilities brought by these developments. At the same time, new religions have provided new meanings, new possibilities and new resources.

The second and third sets of chapters concern peoples situated to the north-east of the Omo Valley, on hillsides and mountainsides that rise out of the Rift Valley. These people include Cushitic groups like the Burji (Amborn, Kellner), the Konso (Amborn, Watson) and the Guji-Oromo (Taddesse Berisso), as well as the Gamo (Tadesse Wolde Gossa), an Omotic group who live on the mountains around Arba Minch town. Also included here are the Gumuz (Wolde-Selassie), who live much further north. They speak a Nilo-Saharan language (the Nilotic languages spoken by the peoples of the Lower Omo, discussed above, are another subgroup of this language family). The experiences of the Gumuz, described by Wolde-Selassie, show certain similarities to those of the Gamo: they have experienced the steady encroachment of their livelihoods by the state since its first expansion at the end of the nineteenth century. This encroachment has been the dominant feature structuring their sense of who they are and their relations with others in the area.

The Konso and the Burji have been described by anthropologists of the Frobenius school (who studied this area in the middle of the last century) as being part of the 'Burji-Konso' cluster (Amborn 1989: 71; Kellner, Amborn, Volume I). This classification was based on similar attributes: language, social organization, symbolism and livelihood. There may be some merit in the idea of a Burji-Konso cluster, as undoubtedly there are many similarities between these people, but this classification plays down some of the commonalities that exist with their other neighbours, for example, with the Boran, who border them to the south-east (Schlee, Volume I). In addition, the Konso, the Burji and the Boran are all classified, linguistically as Cushitic peoples and share aspects of culture. Detailed ethnographic work on Omotic peoples to the north-west, such as the Maale (Donham 1985, 1990) and the Gamo (Tadesse Wolde Gossa, Volume I), shows that there are further similarities in the cultural institutions (for example, in their sacred forms of authority) between these Cushitic peoples and others. The broad linguistic or cultural categorization of people in this area (and others) should not be applied too rigidly; nor should it make the examination blind to the many forms of identification and alliance that cross these conceptual boundaries.

Most of the people in this group have inhabited the same land for centuries and practise intensive forms of agriculture. The Konso and the Burji have terraced their land and cultivate grain-based crops using hand tools. Their investment of labour on the fields in the form of terraces promotes the permanent cultivation of the land, which is made possible in this area of unreliable rainfall only through combining various soil and water conservation techniques. These include the application of manure to the terraces, and the combination of various crops in innovative and flexible ways (Watson 2004; Amborn, Volume I). The Gamo, who live on mountains that rise to a height of 3,000 masl, cultivate enset (*Ensete ventricosum*), or 'false banana', and use its root as their staple food. Tadesse Wolde Gossa's chapter reminds us that enset takes at least five years to mature, sometimes longer at higher altitudes, and that such a livelihood also produces particular ties to the land.

The Gamo, Burji and Konso have lived by exchanging their surplus agricultural crops with the surrounding pastoralists, such as the Boran, Guji and Tsamako. The Burji and the Konso also have specialist craftspeople, who produce cloth, pots, leather, tools and ritual items that served the towns and their surrounding communities with goods before commercially manufactured goods became available with the expansion of national and international markets. Ritual objects used by pastoralists such as the Boran are still obtained from Konso smiths (see Amborn, Volume I).

The existence of these settled agriculturalists on land rising out of the Rift Valley stands in marked difference from the lives of the agro-pastoralists of the Lower Omo and the pastoralists on the Ethiopia-Kenya frontier. The settled agriculturalists are viewed with disdain by their neighbouring pastoralists; as Abbink explains, they are seen as 'short-statured, toiling highlanders'. The pastoralists' and agro-pastoralists' lives are marked by mobility and movement, as a result of insecurity or food shortage, or simply in search of better pastures. Although places have symbolic importance for pastoralists and agro-pastoralists and are central to the construction of identity, their significant places are often visited in pilgrimage or in particular patterns of movement (see Schlee and Wood, Volume I). In contrast, the Burji, Konso and Gamo have highly sedentary lives. Their permanence on the land has given them a strong connection to it, and it is central to their processes of identification and to their senses of self. A recent publication has called for better understandings of the relationships that exist between highlanders and lowlanders and of the significance of these interrelations (Pankhurst and Piguet 2004). This collection is a contribution to such a project.

The settled nature of existence of these agriculturalists has allowed people to invest in institutions that enable particular kinds of identification and alliance building. Amborn argues that the flexibility and diversity of techniques employed by the Burji in their agriculture have made them versatile in their construction of extended networks; this in turn has allowed them to take advantage of new opportunities provided by the expansion of the market in and between Kenya and Ethiopia. The Konso *fuld'o* – the network of craftspeople, which is indigenous and formalized – can be seen as a highly developed example of these institutionalized kinds of indigenous networks. Amborn traces the history of the network and explores the way in which it was a product of the changing market and of the creativity of particular charismatic individuals. Kellner, also working among the Burji, explores the way in which oral histories and myths serve as a kind of institution both for strengthening and for undermining alliances between and within groups in the region.

Tadesse Wolde Gossa looks at the impact of changes brought about by the state on the intimate connection between the Gamo people and their land. Much of their land has been alienated from them as a result of the development of the state over the last century. His study is enriched by comparing the Gamo with the Hor (Arbore), another 'Cushitic' group with whom he has worked, who are situated in the lowlands between the Burji-Konso cluster and the Omo Valley. Wolde-Selassie Abbute's chapter is also included in this section as the Gumuz, whom he discusses, have been similarly disadvantaged, as their territory has been encroached upon by others. In this case, the relative fixity of the people's livelihood practices comes into play again, but in this case the Gumuz are disadvantaged because of their shifting cultivation. Immigrant farmers, who practise plough cultivation, have been allocated land in state resettlement programmes that the Gumuz sometimes use while it is fallow. This encroachment has made Gumuz livelihoods more difficult, and it has also contributed to the increasingly hostile nature of relations between them and their neighbours.

Watson's work also looks at the changing nature of self and identification on the ground, exploring the impact of the Ethiopian state's new ethno-federal process with its emphasis on culture and nationality. Here too the intensive agriculture has played its part, as it has been used by the Konso people to market themselves as a unique culture and nationality. The policies have, however, been appropriated differently by people in the administration in towns from those in different neighbouring villages. In both locations, the decentralization process has brought about a more self-conscious debate about the culture and the role that it should play in everyday life, in constructing relations to others and in gaining access to resources from the state. In Taddesse Berisso's chapter the ethno-federal process is seen to have an equally profound effect. The Guji have developed a heightened sense of political consciousness, in which they emphasize their Oromoness, a form of identification of wider scope, shared with the Arsi and the Boran. It is argued that Oromo ethnic nationalism has developed because of the work of groups like the Oromo Liberation Front (OLF) and the Oromo People's Democratic Organization (OPDO) and as a consequence of the new regional boundaries. The drawing of the new ethnically defined regional state boundaries and the policy of using local languages as the medium of education in elementary schools have provoked emotional disputes and conflict. In this post-1991 conflict, groups that were once considered natural enemies have become allies, and vice versa. In this description of the reversal of group alliances, Taddesse Berisso's chapter provides a useful bridge between this section and the next.

The fourth and final set of chapters in this volume describes the experiences of cattle and camel pastoralists whose territories have been cut across by the establishment of the border between Ethiopia and Kenya. Wood writes about this region of Kenya as a 'northern frontier', where 'the British simply sealed off the place, left people more or less where they were' (Wood, Volume I). The Ethiopian state policies towards the region were not very different. Schlee's and Wood's chapters are closely related: Wood's concerns the Gabra people, while Schlee's traces the shifts that have taken place in the relations between the Gabra, Boran and Garre. Schlee's work shows how each one of these groups' senses of themselves is intertwined with the actions and expectations of the others. In addition, he shows how ideas about these different groups are used by the media and politicians, and the way in which these discourses rebound on relations between groups at the regional level. He also looks at the impact of religious change, but, whereas many of the chapters in these volumes explore conversion to Christianity, which is prominent among the people they have studied, in this case, conversion to Islam is more important. As with Christianity, conversion to Islam transforms processes of identification, and systems of meaning and practice, but, not surprisingly at the current present time, it is also highly politicized; the processes become the subject and object of lively discourses at regional, national and international levels, which have an impact on individual lives.

Wood's chapter returns to the theme of fixity and mobility in everyday life. He argues that, for the Gabra camel herders in the dry lands of northern Kenya, to be Gabra is to be moving. The everyday patterns of movement lead to particular relations to land and to different conceptions of space and place from those of sedentary communities. Spatial metaphors such as 'boundaries', which have dominated anthropological theorizing about ethnic groups, are based, Wood argues, on metaphors imported from sedentary cultural contexts. Wood suggests that, in order to understand processes of identification and alliance formation among nomadic populations, it would be more useful to build on metaphors that are employed by the people themselves. The Gabra, for example, use the metaphor of the 'gateway' or 'doorway', which, as well as being passed through, can also be closed. Employing such metaphors might allow us to construct models of mobile ethnicities that are closer to those lived and felt on the ground.

Together these chapters provide a portrait of the way in which identifications and alliances are produced through a combination of interwoven factors and processes. These include the mobility and fixity of people's lives, which may in turn be connected to their environment and to the rhythms of the livelihood strategies they have developed over time. They also include the lived and bodily experiences and expressions of historical enmities and friendships between peoples and groups. Identifications and alliances are embodied in body scarifications and styles of dress, which are combined with physical givens, such as pigmentation. These are then loaded with meaning and used to categorize and position people. Kinship and marriage relations are constructed in ways that facilitate certain inter-group relations and discourage others. On top of these, the role of the state, modernity, Christianity and Islam, globalization and development are creating new forms of group identification and alliance and reworking the old. Borders, regional and national, are seen here to have a particularly profound effect. In addition, the chapters demonstrate that there are different forms of alliance, some of which are more inclusive and inter-ethnic than others. There are also different kinds of borders: some are permeable and others powerful. In each case, the particular process needs to be investigated; its nature and influence cannot be assumed. The chapters together point to the way in which changing identifications and alliances in this region can only be achieved through unpicking these hybrid processes, at once material and social, political and cultural, operating at different and interlinked scales.

Time: Political Histories

The chapters in this volume describe the experiences of people living in modern Ethiopia and Kenya. The histories of the people in each of these countries are different, but over the last hundred years all have experienced the increasing reach of the modern state. These states have shared the same challenge of administering large areas, often with relatively few resources. Many have relied on ideas and units of ethnicity to help them with this project, which itself has contributed to the crystallization of, even the invention of, ethnic identities. It has often reduced the mobility of people and fixed boundaries between them. In all areas, incorporation into the state and state policies have had some influence on the alliances and processes of identification.

Ethiopia

Ethiopia is seen as unique in the region for having resisted the long-term foreign colonialism experienced by the other countries. The people discussed in this volume live mainly in the borderlands of modern Ethiopia, and were incorporated into the Ethiopian empire-state in the 1890s. Emperor Menelik II, from his new base in Addis Ababa, expanded control over territory at the same time as the European powers were 'scrambling' for control over strategic economic and political territory: his control over the Ethiopian borders was in part a response to the threat of the British, Italians and French, as well as the threat from Egypt and from the Mahdists in Sudan. Although Menelik's armies did not have as much firepower as the European colonists, the process of incorporation was nonetheless often violent and bloody and was always disruptive (see Tadesse Wolde Gossa in this volume for details of the Gamo experience).

In the early decades of the twentieth century, the newly incorporated borderlands were divided into provinces, which were administered by governors, usually members of the nobility who had acquitted themselves well in the battles subjugating the areas in question. These governors had a great deal of autonomy, however, and only had to supply the Emperor with a fixed amount of tribute from his province. Governorships had the potential to provide a great deal of personal wealth through managing the tradable resources in their areas: gold, ivory, civet musk and slaves from the south-west; coffee, gold and ivory from the west. Tribute was also collected directly from the people, through taking livestock or honey or through their labour. Markakis (1974) estimates that this labour – which was called *gabbar* labour – required about one-third of each person's time. The governors and all those who were associated with them – court officials, tribute collectors, soldiers – were allocated *gabbar* to work for them. When a person was doing *gabbar* labour, he or she would often leave his own family and fields and go and stay with those who became thought of as 'northerners' or 'central Ethiopians' for the duration.

The extent of and control over the people and territory varied tremendously. In the hot and malarial Omo Valley, control was nominal; local inhabitants were able to use their superior local knowledge to evade capture and claims to tribute. The northerners used the land for hunting, and made only periodic attempts to gain a more substantial hold over the local population. In regions where people were sedentary and where the land was more fertile, control was stronger. The governors and their entourage engaged in farming and other productive activities.

In all of these border areas, the northerners appointed local people to act as intermediaries and to collect tax. The people chosen for these roles were usually from local notable families and sometimes ritual leaders. In this way, the Ethiopian Empire built on the idea that local customary dignitaries could be harnessed for their own administrative purposes, which inevitably led to some transformations in these indigenous positions. There does not appear to have been a wholesale management of people in terms of ethnic groups, however, or attempts to draw boundaries between them. Provinces tended to be conceptualized pragmatically by the 'exigencies of mountain terrain' (Donham 2002: 5) and determined by the geographical areas that could be divided up and administered from garrison towns.

The experience of this time of empire brought with it certain sets of values that were imposed on other people and were internalized, adapted or rejected in various degrees. This was based on a celebration of northern Orthodox Christian Amhara or Tigrinya-speaking culture, which was considered synonymous with the empire and the apex of an Ethiopian 'civilization'. With the establishment of empire, a movement of administrators, settlers and soldiers into the north began (see Tadesse Wolde Gossa, this volume). For the northerners, the journey from the centre of the empire in Addis Ababa to the remote country of the south – of which the Omo Valley was one of the furthest points – was not only a journey in space; it was simultaneously a journey in what can be described as akin to evolutionary time (Fabian 1983). As the northerners travelled further from the centre of their empire, they placed the people that they encountered further down a scale on which the people were considered to become more 'primitive'. In Ethiopia this categorization was made more powerful because it was also racialized: the further from the centre of empire the northerners went, the more dark-skinned people became. The colonial encounter in Ethiopia was therefore one that opposed the lighter-skinned Orthodox Christian northerners, who made claims to moral, cultural and personal superiority, to the darker-skinned 'others' at the margins, whose morality, culture and personal abilities were derogated as a consequence. The nature of this encounter is encapsulated in the northerners' derogatory term for all the 'black people' living in the marginal areas of empire: *shanqilla*. Although many changes have taken place in subsequent decades and periods of Ethiopian history, the power and legacy of this encounter between the centre and the periphery remain. Both volumes demonstrate that many of the stereotypical ideas and discourses about the 'other' endure (and are similar to strong internalized colonial discourses that exist elsewhere). The movement of northerners into the south has continued throughout the twentieth and into the twenty-first century, as large-scale settlement programmes continue. The identities of many groups discussed in this volume are constructed, in part at least, in relation to this encounter with northerners. One aspect of their identities therefore is as non-northern, non-Amhara, southern or peripheral 'others'.

The period of imperial rule was broken from 1936 to 1941, when the Italians took control over the country. This short period of foreign control differed from many other experiences of European colonization in Africa, in that the Italians' control was limited mainly to the major towns; they never managed to subdue the patriot resistance that fought against them. In Bahru's words, the 'brevity and precariousness [of the Italian occupation] have made it more in the nature of an interlude in the course of modern Ethiopian history' (Bahru Zewde 1991: 166). In the south, the period of rule was noteworthy because the Italians, in theory, outlawed *gabbar* labour, and this, according to Amborn (1988), meant that in some areas they were heralded as liberators. The Italians did, however, still require that the peasantry provide them with porters and labour of various kinds when they needed it: in the minds of many of the people 'liberated' from the northern conquerors, little had changed. The other impact that this period had was that the Italians recruited local people to fight for them, whereas many others joined the bands of patriot resistance fighters, which were widespread across the country. This brought divisions within and across communities, which had long-term impacts.

After the Italians were driven out, Haile Selassie, who had already been crowned Emperor in 1930, returned to the throne. In the post-Italian years, he ushered in a new period of modernization in the country. He encouraged education and industry, established some health care and outlawed tribute collection, replacing it with a centralized tax system. These developments were also related to the new emphasis on Ethiopia as a united and centralized state, run by the Emperor from Addis Ababa, instead of a loosely connected set of fiefs. The new tax system undercut the powers of the provincial governors and administrators, who were loath to give up their privileges for a more bureaucratic and accountable system. As taxation demands of the state increased, many in the periphery found that they still had to pay directly to their 'northern' administrators. Thus this period was often characterized by the double burden of meeting old and new demands simultaneously.

In 1974, Haile Selassie, by this time an old man, was overthrown in a 'creeping coup'. The coup was led by a group of military men who had become increasingly frustrated with the Emperor's style of government. The Emperor championed the modernization of the country, but his rule was still autocratic, highly personal and imperial. Poverty, famine and inequality were widespread, and the Emperor was seen as far from providing effective solutions to the problems; he appeared oblivious to them and to be enjoying the privileges of his position. He was pictured in the media as 'uncaring and wealth-crazed' (Donham 1999: 21). In 1974, public dissatisfaction was so widespread that '[p]ractically every group from prostitutes to lay priests went out on strike' (Ottaway and Ottaway 1978: 4). The Ethiopian student movement campaigned for several years for land reform under the slogan 'land to the tiller', and this also contributed to the end of Haile Selassie's reign.

The group who led the coup came to be known as 'the Derg', from the Amharic word for committee, and included representatives of the army, air force, navy and police. They soon announced that they espoused a form of Ethiopian socialism, and set about implementing policies that would bring about their aims of 'equality; self-reliance; the dignity of labour; the supremacy of the common good; and the indivisibility of Ethiopian unity' (Ottaway and Ottaway 1978: 63). They nationalized all banks and major industries, and, in 1975, implemented a 'thorough and radical' land reform programme (Dessalegn Rahmato 1984): all land was declared the property of the state; those cultivating the land only had usufruct rights; landholdings of more than ten hectares were made illegal; and leasing, borrowing or mortgaging land was forbidden. The previous grass-roots administrators were dismissed and new grass-roots units were set up called Peasants Associations. 'Students' – a category that included university students, schoolteachers and all those in the last two years at school – were sent into the countryside to help to achieve this new vision of the Ethiopian state. In 1978–79, agricultural production quotas were introduced. From 1985, a villagization programme was implemented. By spatially and socially reorganizing village society, it was hoped to modernize and develop the 'peasantry' further.

The top-down, modernizing regime that the Derg represented completed the unification of the Ethiopian state, and stretched the hand of the state more deeply into the lives of the Ethiopian people than any regime had done before (Clapham 2002). The rule of the Derg is also remembered for terror and bloodshed: there was internecine fighting for power between Derg factions, a struggle that was ultimately won by Mengistu Haile Mariam, who became the Derg Chairman and dominated the government until it fell in 1991. Those who were considered to represent a political threat to the government were sought out and killed, culminating in the 'Red Terror' of 1977–78.[2] There was fighting with Somalia in 1977–78 (the 'Ogaden War'), and protracted conflict with insurgencies in Tigray and Eritrea. The state's counter-insurgency measures required that large numbers of young men were forcefully conscripted into the army, many from the peripheries. The Eritrean and Tigrayan resistance ultimately defeated the Derg in 1991.

The post-1991 government is known as the Ethiopian People's Revolutionary Democratic Front (EPRDF), and has been dominated by the Tigrayan People's Liberation Front (TPLF), and led by Meles Zenawi, Chairman of the TPLF and Prime Minister of Ethiopia. Soon after taking power, this government implemented a radical programme that federalized the country along ethnic lines. The Constitution, ratified in 1994, provided each ethnic group, or 'nation, nationality and people' as they are referred to, with the 'unconditional right to self-determination, including the right to secession' (Article 39, Federal Democratic Republic of Ethiopia, 1995: 13). These developments fit well with ideas about human rights, cultural autonomy and post-colonialism current in the global neoliberal order. In Ethiopia, they also have an

---

2. The attacks were mainly against those thought to support MEISON (All-Ethiopia Socialist Movement) and EPRP (Ethiopian People's Revolutionary Party).

older history and resonate with Stalin's policies towards nationalities in the former USSR (Clapham 2002; Watson 2006).

In the new ethnocracy, a vast amount of energy has been spent on debating, drawing and redrawing the lines of the new ethnically defined states within this federal structure. The outcomes of the policy have been mixed. For some, especially the small ethnic groups in the south that never before enjoyed any recognition from the state, it has, for the first time, provided them with a stake in public policymaking (Watson, this volume). For other larger states, it has given them some autonomy. But, overall, experience in the years that have passed since this radical policy was implemented suggests that the whole experiment has had limited success. While espousing the autonomy of the regional state, the central government has maintained tight central control, frequently compromising regional policymakers. Questions have also been raised about the control and distribution of state resources (Clapham 2005). In 2005, Ethiopia witnessed multiparty elections, which were initially described as free and fair, but the release of the election results was delayed and the outcome contested by the ruling government. When people protested, tanks were sent out on the streets, and there were bloodshed and new detentions of people considered to be political dissidents (see Abbink 2006, for a review). These events have led to accusations that the EPRDF's commitment to change and democracy and to giving political voice to people at the grass roots is, at best, only skin-deep.

Economically, the country has, over the last five to ten years, opened up its markets to foreign investment. So far the government has resisted calls to denationalize land, but it has made some compromises towards investors by allowing land to be leased and mortgaged. The conflict from 1998 to 2000 with Eritrea, which seceded from Ethiopia in 1991, was also costly; tensions are still rumbling on today and new conflict is not out of the question. In terms of health, Ethiopia has the lowest recorded level of HIV/AIDS prevalence in the region for which figures are available, but at around 4.4 per cent it is not insignificant. In addition, although there has been a global commitment to making antiretroviral therapy available to the poor, less than 10 per cent of the infected population in the country is estimated to be able to access drugs, and even this estimate seems optimistic.[3] At the time of writing, there are still serious challenges facing the country.

Kenya

In Kenya, historians, archaeologists and anthropologists have all emphasized the fluidity of people in the pre-colonial situation (Sobania 1988a; Spear and Waller 1993): 'They were peoples, not tribes ... There were boundaries between them, and they gave their neighbours different names, but these served to demarcate different environments and different cultures that had grown up in their management, not absolute breaks in political allegiances and economic self-sufficiency. Trade, marriage and patronage knew no confines' (Berman and Lonsdale 1992: 19). The boundaries that existed were fluid,

---

3. UNAIDS/WHO AIDS Epidemic Update: December 2005 (http://www.unaids.org/ epi/2005/doc/EPIupdate2005_pdf_en/Epi05_05_en.pdf, accessed April 2006).

and forms of exchange and mutual support were institutionalized (Sobania 1988a). After the violent conquest by the British administration in the late nineteenth century, the colonial state attempted to make good their investment by extracting labour and resources. Large areas of the best agricultural land were taken over by expatriate settlers, and, in the areas that became known as the 'white highlands', many became displaced. Others' livelihoods and existences became newly restricted to particular regions. The colonial administration understood territories and 'tribal units' as belonging together, and by mapping them they fixed them on the ground, creating a particular 'order' (Berman and Lonsdale 1992).

The areas of the north of Kenya that are most relevant to this book were not given so much attention by the colonial administration. They were seen as expensive to manage and unproductive, but also as an important buffer zone between Kenya and the Ethiopians and the Italians and the French. As in some other regions, the history of people and their relations to the state were shaped by the combined impact of bovine pleural pneumonia, rinderpest and smallpox, which devastated herds and lives at the end of the nineteenth century. In the dry northern territories, herders, who had previously ranged over large territories, exploiting different grasslands in wet and dry seasons, consolidated themselves and their remaining herds in smaller areas of land. Sobania's work on Rendille and Ariaal herders to the east of Lake Turkana illustrates how the colonial administration encountered this situation at the end of the nineteenth century and took it to represent the 'traditional' and historic grazing pattern (Sobania 1988a). They used what they found on the ground to delineate ethnically defined rights to pasture, and through this the much-reduced grazing pattern became official.

Sobania (1988a) describes how the limited grazing of the Rendille and the Ariaal was further exacerbated when the British encouraged Boran and Gabra, whose territory had been cut by the Ethiopia-Kenya state border, to use land that the Rendille had previously accessed in dry periods. The British established 'tribal grazing areas' with the aim of administering the pastoralists, which included taxing herders and reducing raiding. The result was the limitation of herders to restricted and fixed grazing lands, which contributed to environmental degradation of the grasslands. It also contributed to an undermining of the connections of mutual aid and exchange that had previously existed between groups, and led ultimately to increased, rather than reduced, levels of raiding and conflict. Development initiatives in the colonial and post-colonial (1963 onwards) state included the development of towns (in which barter trade was banned) and the introduction of health care and formal education (largely by missionaries) and water development projects. The water development projects of the 1950s and 1960s led to the intensification of use of grasslands around new water points and an associated environmental degradation and undermining of livelihoods; the development of towns and services led many to settle around towns. Some herders have benefited from the new opportunities, but the accounts of this period repeat similar themes: reduced levels of mobility; stronger boundaries between groups; heightened notions of ethnicity or whichever level of identification became relevant for the allocation of territory; reduced levels of inter-group interactions (particularly of a dispersed, informal, individual-to-individual form); environmental degradation; and poorer livelihoods.

In the 1960s the Northern Frontier District (NFD), as it was then known, became the site of a fierce guerrilla conflict between pastoralists and the state. These '*shifta* [bandit] wars' were brought about as, on independence, some northern groups, particularly those that were Muslim and Somali-speaking, preferred to become part of a Greater Somalia. Such a move was resisted by Jomo Kenyatta's government. Scholars argue that the diverse pastoralist groups in the north were viewed as guilty by association: 'all Borana and Somali were lumped together and regarded with suspicion by the state. Soon, any person from the NFD region was commonly referred to as a *shifta*, and this display of hostility on the part of the state (and many Kenyans) stemmed directly from the general belief that all "northerners" supported secession to Somalia' (Arero 2007: 297). The northern pastoralists were punished for their rebellion: 'vast numbers of animals were confiscated or slaughtered, partly in order to deny transport to the guerrillas, and much of the conflict was confined to a few population centres, where an underclass of destitutes developed' (de Waal 1997: 40). The policies of the *shifta* war began a systematic attack on the pastoralist way of life, an attack that has continued to some extent to this day. The groups in the north have become significantly pauperized as a result, unable to be resilient in the face of droughts experienced in recent decades (Baxter 1993; de Waal 1997).

Despite these serious problems in the north, Kenya was seen until the 1990s as a beacon of stability and economic success in the region. In the 1990s, however, Kenya no longer had the strategic importance it had enjoyed during the cold war period. As in many African countries, livelihoods suffered as a consequence of economic crisis, increased debt repayments and structural adjustment programmes.

In development policy circles, the 'good governance' agenda rose to prominence, and indigenous pro-democracy campaigners became increasingly vocal, inspired by what they saw in Eastern Europe and the success of anti-apartheid movements in Namibia and South Africa. Under domestic and foreign donor pressure, the Moi government finally agreed in 1991 to shift to a multiparty electoral system. The move was followed by the worst fighting in Kenya since independence. Despite many alliances that crossed ethnic lines, the conflict was viewed as ethnic, and there was some truth to this: much of the conflict was inflamed by old grievances that had originated in the policies of the colonial and post-colonial state, which had allocated land to people by ethnicity. Other grievances resulted from allocations of land that neglected ethnicity. In many cases, for example, land in the Rift Valley was, after the departure of white large-scale landowners, not given back to the people from which it had been confiscated in the early decades of colonization,[4] but to Kikuyu who had followed the white settlers as labourers and squatters. Large tracts of land were also bought or 'grabbed' by post-colonial Kikuyu elites. A government report confirmed what many observers believed, however, that the conflict had been instigated by the ruling government in order to fulfil Moi's prophecy that multiparty politics would lead to 'tribal clashes'. The conflict also

---

4. Much land was needed to settle demobilized British military officers. The original owners of the land often found themselves displaced for the benefit of those for whom they had fought as soldiers against the Germans in Tanganyika or for whom they had served as porters in that war.

displaced many from their home areas and this provided a form of gerrymandering before the first multiparty elections (Haugerud 1995).

The conflict was also a result of the withdrawal of state services and a decline in livelihoods that accompanied the structural adjustment programmes and the neoliberal policy agenda of the 1990s. In the absence of the functioning state, many turned to patron-client relations for support. The result was the further personalization and informalization of politics, as well as a further emphasis on clan-based and ethnic networks and identities. From the end of the Moi regime in 2002, there were great hopes for the new government under Mwai Kibaki. But the recent fiasco of the 2007 Kenyan election and the widespread conflict that has followed it have dashed these hopes. The conflicts have had a strong ethnic dimension and demonstrate again the importance of understanding the origin, form and impact of identity politics. The cases presented in these volumes show, however, that identity politics in Africa can sometimes be ethnic, but they also have many other dimensions and forms. On the other hand, the ethnicization of politics in Kenya is not a recent phenomenon. It has its roots in ethnic territoriality being made the basis of the colonial administrative order (Schlee 1994a) and has moved on progressively through the Kenyatta and Moi periods of presidency. Schlee's contribution to this collection shows how in the late Moi period ethnicity, from being a matter of 'unofficial' history and conspiracy theory, more and more moved into the open to become the overtly stated reason for political claims and for the exclusion of others.

Many of the hopes fixed on the change of government in 2002 have also collapsed as the new government has been accused of similar levels of corruption and ineffectiveness to those of their predecessors. The impact of the HIV/AIDS pandemic should not be forgotten as an important contributing factor to this history. HIV infection prevalence rates reached a peak of 10 per cent in adults in the 1990s, and have declined since then to approximately 7 per cent in 2003.[5] Despite the evidence for positive trends, the HIV pandemic continues to present a challenge to livelihoods and governments.

A set of volumes like the present one, which aims at being scholarly and to have some depth of analysis, showing major trends, patterns of alliances and the ways some of the cases under study are interconnected, takes some time to prepare. It can never be up to date in the way other media (such as the press or the Internet) can. It may not include the latest initiative for a peace process or the latest change of government in one place or another. Even if it were up to date in this sense, it could only remain so for a while.

What we aim at is to identify some forces at work over a longer period and to show some lasting or recurrent configurations into which things fall again and again. If we have succeeded in making the structuring principles behind the changing identifications and alliances in North-East Africa more clearly visible than they have been so far, these volumes will not become obsolete soon. Should major political or social changes occur that also affect these underlying patterns, these changes would not render our account obsolete but 'historical'. Or so we hope.

---

5. UNAIDS figures (http://www.unaids.org/epi/2005/doc/report_pdf.asp, accessed April 2006).

Part I
Identification and Insecurity in the Lower Omo Valley

Chapter 1

The Fate of the Suri: Conflict and Group Tension on the South-West Ethiopian Frontier[1]

Jon Abbink

Introduction

The Suri people in south-west Ethiopia are one of the minority groups (numbering about 28,000 people) that under Ethiopian electoral law have automatic representation in the Ethiopian House of People's Representatives because they do not form a unit large enough for an electoral constituency. Since 1995 the seat for the Suri had been occupied by Guldu Tsedeke, a promising young Suri man and well accepted among the people themselves. He was re-elected with a large majority in May 2000 as an independent candidate (i.e. not affiliated with the ruling party). In early January 2002, this MP was killed in a shoot-out in Suri country, when he had gone with a few friends to negotiate with a wanted Suri murder suspect. This was also the man who shot him.

The death of Guldu Tsedeke, a major Suri figure and the first one to more or less successfully mediate politically between the modern Ethiopian state and Suri society, is a starting point of my chapter discussing group tensions and conflict in the poly-ethnic Ethiopian Southwest. The case – to which there was no political background – is indicative of what I see as a serious crisis in a society that has not yet come to terms with itself – or with the problems posed by the era of authoritarian state formation and of locally manifested globalization. Here I focus not only on the state impact on local societies but also on the ramifications of changing alliances and identifications among and within the groups themselves. A debate on such changing realignments of small and 'peripheral' groups in 'ethno-federal' Ethiopia is timely and challenging (see James et al. 2002 for a recent collection on this topic). In the southern Ethiopian regional state there are, besides a few large groups like the Sidama, Wolayta, Gamo and Gurage, dozens of such numerically smaller groups that

---

1. The author and editors wish to express their gratitude to the Editorial Board of the *Journal of Eastern African Studies*, and to the Taylor and Francis Group (London), for their permission to use a paper published in the journal as the basis for this chapter.

form an essential part of the dynamics of the multi-ethnic heritage and current politics of Ethiopia (see Abbink 1998b).

In the past fifteen years or so, a notable increase of violent local group conflicts was evident in south-west Ethiopia, as incidents were recorded not only between certain ethnic groups and the state (as with the Hadiyya, the Maale and the Me'en), but also among and between groups, for instance Anywaa ('Anyuak') and Nuer, Gedeo and Guji, Bodi and Dime, Mursi and Ari, Boran and Hamar, to name but a few. A number of these conflicts – which left many hundreds of people dead – relate in part to customary patterns of rivalry and raiding that existed in the past (see Fukui 1979). This link, however, is always made by government administrators and power holders to excuse their own inaction or inability to de-escalate the problems. It does not provide an explanatory argument for current fighting and the new forms it takes. First, the nature of conflict has significantly changed: more arms are involved, more people are being killed and the rules of engagement are changing and those of reconciliation deteriorating (see Young 1999; Abbink 2000c; Tronvoll 2001). In addition, the army or police in Ethiopia are rarely sent to contain the problem, and people know this, so the fighting escalates. Thirdly, partly due to post-1991 state policies, there is an essential difference in the conceptualization of and response to conflict. Current group differences and disputes are now referred to, including by many local people, as being 'ethnic'. This line of thinking is predictable and easy to resort to but, as a kind of cultural-essentialist argument, it is not convincing. People seem to fall unwittingly into the trap of adopting the state-sponsored discourse on ethnicity. While not denying the great importance of ethnic belonging and its social, emotional and cognitive roots in the *habitus* of people, I contend that it neither constitutes people's entire social identity nor provides a full explanation of their behaviour and their choices.[2]

The basic theoretical question is always how and why conflicts of interest come to be seen exclusively in terms of 'ethnic antagonism', and whether it is helpful to act on such a perception. Cultural-essentialist views have been discredited in anthropology for quite some time (for a good historical discussion, see Kuper 1999). So this chapter ties in with a theoretical debate on the dynamics of ethnicity and conflict, a field where several theoretical traditions dominate: the resource competition theory (see e.g., Markakis 1998; Homer-Dixon 1999); symbolic theories referring to pre-existing cultural differences (the Geertzian approach in anthropology); and perhaps theories of hegemonism as a sociocultural phenomenon. I would opt for the last line of thought, but would plead for extending it with what I would call humiliation theory.[3] Here, prestige, honour and identity issues are seen as cultural or social constructs, which, though based on material differences and

---

2. Which await investigation and evaluation in a comprehensive, critical manner, despite the spate of papers and articles published in recent years. The emphasis has been on formal political developments and policies, not on their effects on and 'appropriations' by ordinary people.

3. There is a growing body of literature on this theme. But a systematic theory linking psychological, social-structural and cognitive aspects is still underdeveloped. For a very interesting study, see Miller 1993.

conflicts of interest, are the cultural-psychological models on which humans act and which become engaged in inter-group relations. Often images of stigma are involved: perceptions of low status that structurally attach to certain communities (for a pioneer study, see Goffman 1963). The above models of and for action are related to questions of belonging, autochthony and survival, and as such are obviously also important in the dealings of local groups with the state. Especially in this context, where a new hegemonic state project is in full swing in southern Ethiopia, there is a need for a systematic analysis of the construction and dialectics of humiliation in the settings where it is produced.[4]

Among the disturbing local conflicts in southern Ethiopia one can now also reckon those between the Suri and their neighbours. In these conflicts, no political-ideological agenda is evident. They are about survival, resources, local rivalries and prestige. In this intricate dialectic, seen from the Suri viewpoint, there are several categories of problematic outsiders with whom they have differences. First, the Suri distinguish between the Dizi, the Me'en and Ethiopian government representatives, and a second category of people whom they consider to be their 'real enemies', like the Nyangatom, the Toposa (in Sudan) and the Anywaa. Quite another third category includes people related to them through language, culture, marriage bonds and ritual relations: the Mursi and the Baale (which is not to say that they don't have problems with them).

In what follows, I briefly discuss the various levels of conflict and see whether there is a process of identity (re)definition and of (re)alignments going on, and whether there are prospects for a more peaceful development, whereby issues of sharing – of resources, peacemaking procedures, ideas of development – can be encouraged. On this basis, it is possible to venture an answer concerning the future identity and patterns of alliance of minority groups like the Suri. The question is urgent because people are being threatened, wounded or killed, almost on a daily basis, and there is a deep feeling of insecurity in the Maji area.

Suri Society and Economy

The Suri live in a sensitive border area of Ethiopia and Sudan, which has been a major venue for arms, cattle, slave and ivory trade in the past (see Garretson 1986). Even in recent years, Suri have purchased or traded arms from across the Sudanese border, and Ethiopian lands are subject to frequent raids from Sudanese ethnic groups like the Toposa, who wreak havoc among Suri cattle herds and have forced them to migrate to the north-east. The presence of an Ethiopian *Fet'no Derash* army contingent – a 'rapid reaction force' of about fifty to sixty soldiers – dispatched by the central government a couple of years ago, ostensibly to 'protect' Suri from attacks from neighbouring troops and keep the peace, has not made much impact. It seems more concerned with gathering intelligence and keeping the Suri in check, rather than moderating the actions of foreign invaders.

---

4. This is not to say that people who are 'humiliated' are always right in the factual or moral sense; we are analyzing the social fact of people reacting in such terms to intensifying contacts, challenges, conflict, and hegemonic state policies.

The Suri are an ethnic formation that, according to their oral tradition, emerged some 250 years ago in the Sudan-Ethiopia border area. Historically, the Suri were located around the Shulugui mountain (also known as Naita) on the border with Sudan, an area now occupied by the Nyangatom. Here their most important ritual places are located: the burial sites of their ritual leaders (the *komorus*) and the places of initiation of age-sets (done about every twenty-five to thirty years). Suri also used to get several vital ritual materials (plants, ritual paint, coloured stone) from this area. The Suri see this place as the 'stomach' (in Suri: *kyengo*) of their country; it forms the core centre of their cultural space, and now is also a *lieu de mémoire*. Their area of settlement is now about fifty to sixty kilometres to the north, close to the Dizi mountains and the western Akobo valley (see Map 1.1).

Their society is an agro-pastoral adaptation to a savannah lowland area of insecure rainfall. Suri have mixed origins, with ancestry from neighbouring groups such as the Dizi, Me'en and Baale and perhaps others in Sudan (Murle). The two subgroups of the Suri – called Tirmaga and Chai – again have diverging origin stories. The language of the Suri is a Nilo-Saharan one ('Surmic' subgroup) and very different from that of most of their neighbours. (Only Baale and Mursi have a similar language.) The Suri are also historically related to the Didinga, Narim and Murle in Sudan.

Suri form a kinship-ordered society, with an important role for patrilineal clans (for marriage exchange, ritual functions). Political authority is vested in a senior age grade of elders, and in a ritual leader without executive power, the *komoru*, a function embodying values of community peace and communication with the supernatural. Territorial organization is in the form of villages and of trans-clan herding units with cattle camps. Suri economic activities are rain-dependent cultivation, panning and trading of alluvial gold and especially livestock herding. The management and expansion of cattle, which are individually owned (although lineage agnates have a claim to it as well) but herded collectively, are the generative social mechanism and ideal in their society, explaining much of Suri behaviour. Cattle take a prominent place in daily life as a medium of exchange, as well as in cultural representations. Historically, the Suri have always practised cultivation and herding as complementary activities. In addition, they practised hunting and gathering.[5] They also trade, on a small scale, livestock, pottery, gold and game products with the people around them, including the highland villagers. They are therefore partly dependent on contacts with these neighbouring groups, more than they would like to admit.

In a cultural sense, Suri are acutely aware of their being different from other groups. They also cherish their virtually autonomous situation in a marginal border area. They never felt stigmatized or inferior. Their cattle-herding way of life is preferred by them above what they see as the dull and toiling farming life of their highland neighbours like the Dizi and the villagers in Maji. There is an enduring resistance among the Suri towards becoming settled peasant cultivators as the Ethiopian government would like to see. Suri cherish egalitarianism and personal

5. The Suri until recently intensively hunted in the nearby Omo National Park (buffalo, hartebeest, giraffe, antelopes; some species, such as elephant and rhino, have disappeared).

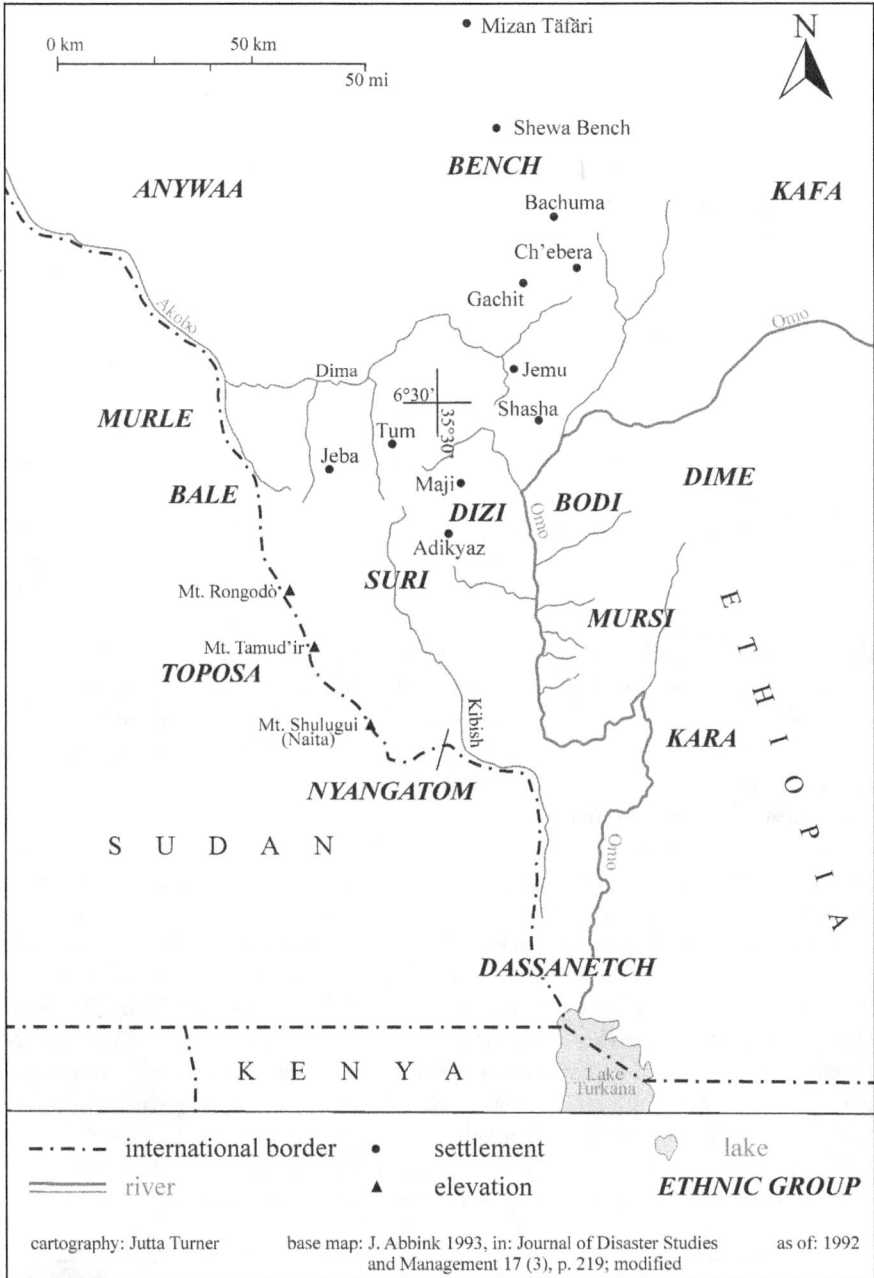

0 km   50 km

50 mi

N

• Mizan Täfäri

ANYWAA

BENCH

• Shewa Bench

KAFA

Bachuma
•

Ch'ebera
•

Gachit
•

Dima

6°30'

• Jemu

Shasha
•

MURLE

Tum
•

Jeba
•

Maji •

BALE

DIZI

BODI

DIME

Adikyaz
•

SURI

Mt. Rongodò ▲

MURSI

Mt. Tamud'ir ▲

TOPOSA

E

Mt. Shulugui ▲
(Naita)

Kibish

KARA

T

NYANGATOM

H

S   U   D   A   N

Omo

I

O

DASSANETCH

P

I

A

K   E   N   Y   A

Lake
Turkana

— · — · — international border   •   settlement   lake

═══════ river   ▲   elevation   ETHNIC GROUP

cartography: Jutta Turner   base map: J. Abbink 1993, in: Journal of Disaster Studies   as of: 1992
and Management 17 (3), p. 219; modified

Map 1.1 Approximate location of the Suri and other groups in the Omo and Akobo valleys.

independence. Women are prominent in social life and can own cattle and small stock, but never participate in herding.

There is a discourse of 'culture difference'[6] between Suri on the one hand and Dizi and other highlanders on the other, which has become more influential in the past years. Suri express disdain for the 'short-statured, toiling highlanders' and for the absence of any aesthetic body culture among them. They see no charm in a lifestyle without ceremonial duelling and, above all, without cattle and the entire cultural complex related to it (naming, cattle songs, dietary customs, ceremonial initiation, etc.). The cultural representations involved here also constitute Suri notions of personal dignity and achievement.

The Suri always show strong group pride and as a rule shun the highland areas and state control as much as possible. They have some respect for the Me'en people and for the Nyangatom, although the latter are their traditional enemies. The greatest contrast and antagonism that they experience are with the agents of the state. This dates back to the moment of their incorporation into Ethiopia in the early twentieth century. In the imperial era they were seen as roaming nomads in faraway border areas, without much interest to the state. Under the Derg regime this perception did not change much, despite the new revolutionary rhetoric of equal rights, as they were considered part of a 'primitive-communal' society with many 'harmful customs' and a lack of education and knowledge. The actual contacts with the government were not conducive to development and rapprochement. The last army post in the Suri area was abandoned in 1988; the two primary schools were closed shortly after. In the later years of the Derg, the Suri became embroiled in conflicts of various kinds with virtually all of their neighbours: Me'en, Dizi, Anywaa and Nyangatom. There was only an uneasy alliance, or at least tolerance, with the Mursi on their eastern border.

Under the EPRDF (*Ehadig*) government, the Suri were redefined as a minority in need of development and education, but continued to be seen as a group with 'harmful customs' and a violent record. Their agro-pastoral way of life was considered to be inefficient and backward. Suri at present have been given their own *woreda* (district) within the Bench-Maji Zone, and virtually no non-Suri lives within its borders. Since 1994, there is a local 'Surma Council' (*Surma Mikir Bet*) with eleven Suri members, with a changing membership. The places on this council are coveted because of the government salary they bring. The *woreda* has some government offices in the small village of Kibish, near an old airstrip dating from the late 1960s when several foreign missionaries were living there (see Tornay, this volume). Non-Suri don't like to serve in this Zone and especially in the *woreda*, and it is not without risks either. In the past five years, several police officers, teachers and

---

6. The term 'culture', as relevant in the local discourse on group relations and enmity in the Maji area, refers to nothing else but the socially constructed and inherited repertoires of difference – in lifestyle, cognition and values of honour, identity or dignity – between human groups. (Locally, in Amharic, people talk of *dänb*, not *bahil. Dänb* means 'the rules', 'the customary ways of doing things'; *bahil* is song, dance, theatre, etc.: the non-problematic, folkloristic culture.)

agricultural agents have been killed in shoot-outs and brawls in the Suri area, not to speak of local Suri, Me'en and Dizi people. Transport facilities are bad, the climate hot and malarial, and remuneration is not noticeably higher than elsewhere in the country. The only attraction is that it facilitates (illegal) trade in alluvial gold (sold in Addis Ababa), which the Suri pan from local rivers and sell locally to outsiders.

In the last decade, the groups in the Maji area were more drawn together in a common political arena, confronted with a new regional policy and a formal recognition of 'ethnic rights' by the new government. But this has led neither to visible socio-economic improvements nor to a more peaceful settlement of disputes. In some cases even the appeal of people to their own presumed ethnic identity and rights has tended to reinforce conflict, and in some respects becoming 'too equal' with others has generated more problems itself.

External Conflicts

In southern Ethiopia, the Suri people have an exceptional place. Living in a remote border area as largely autonomous agro-pastoralists, they have built up a reputation of trouble and violence among their neighbours. Suri feel that they are masters in their own land – the lowlands south-west of Maji town, extending into Sudan – and they recognize no overlords. They know that the Ethiopian government and its army are powerful, but also that it is difficult for the state to control the Suri in their own remote lowland region. Not all Suri pay taxes, nor is it possible for the government to maintain a strict monopoly on the use of violence. The Suri self-perception of independence and autonomy is the basis of their vigilant self-defence and of their preparedness to use force. In this, they are not exceptional compared with many other East African pastoral groups. The nature of the pastoral economy almost by definition brings them into conflict with neighbouring groups, notably the Nyangatom and Toposa. But in the past there used to be social contacts with these groups, bond partnerships, some trade and a code of fighting.[7] All this is now rare: Nyangatom and Suri do not meet, except for some of their young (zonal) leaders in Awasa (the capital of the Southern Region) when they are called there to attend policy meetings. These young leaders, who are drawn into the agenda of the ruling party, do not have much influence yet on the rank and file of their peoples, although in recent years the zonal and regional authorities have been able to call peace meetings (2006–2007).

At present, group relations in the Maji area are still marked by tension and violence. This holds for: (1) Suri – Nyangatom (aim: territory, guns, cattle); (2) Suri – Toposa (aim: territory, cattle); (3) Suri – Anywaa (aim: gold, money);[8] (4)

---

7. Which included the giving of coded warnings before an impending cattle raid, no burning of pasture, no poisoning of wells, no killing of cattle, no raping and killing of women.

8. Anywaa were not 'traditional enemies' of the Suri. Incidents with the Anywaa have occurred over approximately fifteen years in gold-panning places on the northern fringe of the Suri area and also in Dima, a frontier town on the River Akobo where the gold is sold. Since 2000, the town is part of the Gambela Regional State, where Anywaa are powerful. In Dima, visiting Suri are frequently robbed of their gold and money. Anywaa, because of their alleged deviousness, are seen by many Suri as their worst enemy.

Suri – Dizi (aim: territory, cattle, girls, clothes, money). To a lesser extent the Suri have problems with the Me'en people, who live on the escarpment east of the Akobo valley and Maji (see Map 1.1). The aim here is also to get cattle, guns or girls.

In the conflicts with these groups there are obvious material interests (see above), but also immaterial ones like 'prestige' and fighting feats to attain 'status' within their own peer group. I do not elaborate on this, but it is an element not to be neglected in a complete analysis of Suri (and other groups') violent performance. In their conflicts with Me'en, villagers and especially Dizi, the aim is also to intimidate. In their confrontations with the Toposa, Nyangatom and often also Anywaa, the Suri are usually the losers. I shall comment on two of the most important conflictual relations: between Suri and Nyangatom and Toposa; and between Suri and Dizi. I shall also discuss in more detail the relations between the Suri and the state.

Suri and Nyangatom and Toposa

The Suri call both the Nyangatom and the Toposa 'Bume'. These are two allied peoples and speak the same Para-Nilotic language.[9] Since the mid-1980s, they have been encroaching on Suri lands. Nyangatom have driven the Suri out of their core area at the Shulugui (Naita) and the T'amudir hills and usurped their best pastures and waterholes.[10] Before the early 1980s, the Nyangatom (some 13,000) were at the receiving end of Suri violence. The Toposa (about 75,000 people) were located further west in Sudan and, before the 1980s, were not much engaged with the Suri. The migration of Kenyan Turkana up north, as well as the chaos of the Sudanese civil war, has pushed the two groups into Suri country. In the early 1990s, the Toposa obtained a regional edge as they were formed into a so-called 'tribal militia' and were well armed by the Sudanese government. Toposa raids occurred almost every one or two months in Suri country. Due to the change in alliances, there is now a pressure on Suri 'resources' related to the pastoral economy that was never there before. But it is not only resource pressure that causes problems for the Suri way of life. The Chai Suri (the largest group) have been robbed of their prime ritual places around Shulugui (see above), and also feel thwarted and impoverished in a cultural sense. Indeed, when talking about the recent past, almost all Suri recall with nostalgia the life at Shulugui and in public debates and private conversations express their frustration at not being able to go back. Only in 2007 a period of *détente* started, with a number of successful peace meetings between the three groups, both in Ethiopia and Sudan. They were held under the auspices of the zonal authorities and seem to have notably reversed enmity, opening up possible new alliances.

---

9. According to Bender (1976: 59) these languages belong to the Eastern Nilotic branch of the East Sudanic subfamily of the Nilo-Saharan language family (Editors' note).

10. The original, rather fluid border between 'Bume' and Suri ran from below Mt. Shulugui to the River Omo, up to Kara country. For about the last ten years it has run from Mt Rongodò to the Dirga Hills, just south-west of the River Omo bend near the Mursi area (see Map 1.1). The Suri have also lost virtually all of their transhumance pastures in Sudanese territory.

Suri and Dizi

For about eighteen years, there has been a notable increase in violent incidents with the Dizi. Dozens of Suri and hundreds of Dizi (a population of some 26,000 people) were killed in armed incidents. Some of the worst of these were a massacre of forty-three Dizi people in the village of Kärsi, in 1990, as revenge for the killing of a Suri man in a fight with a Dizi, and a similar killing in Kolu in 1993, with thirty-five Dizi killed, including the chief of Kolu. The cold-blooded murder by Suri youngsters of three young women later in 1993 was also deeply resented by the Dizi. The same year the Suri raided the big Dizi village of Jeba and killed twenty-three Dizi and three policemen.[11] In recent years, attacks on Adikiaz, Jeba, Kolu and even the outskirts of Tum, the district capital, have left a few dozen people dead. The list is long, and perpetrators are rarely brought to justice.[12]

The increased tensions with the Dizi (and also with the Me'en) are directly related to the breakdown of the Suri-Nyangatom relationship. It seems to be a simple case of displacement of aggression. No longer capable of properly retaliating and regaining their cattle from the 'Bume', the Suri seek replacement from the Dizi. As the Dizi are an ill-armed sedentary population[13] and have some cattle, they are an easy target for Suri raiders. Dizi do not venture into the lowlands in pursuit of the Suri. They are a very frequent victim of both cattle theft, raids on houses and ambushes on the roads.[14] In view of the past relationship between Suri and Dizi this development is surprising and alarming. The Dizi are ancient settled cultivators in this area with a hierarchical chiefly society. Suri arrived later from the south and entered into contact with them. Both Dizi and Suri traditions maintain that their leading families share common descent and cannot intermarry, although the common people could. Historically, both groups had a kind of ritual alliance whereby the Dizi chiefs were recognized as having rainmaking powers. In times of

---

11. Not only Dizi but also the Me'en have been victims of Suri raids, especially after the EPRDF regime made a major effort to disarm them. Me'en are locally seen as the best fighters. In late 2000 the Me'en in the Gesha and Kella areas had had enough of Suri attacks and ambushes and staged a carefully planned and well-organized retaliatory raid, whereby they took about 5,000 Suri cattle: one of the largest raids ever seen in the area. Some months later the Suri randomly attacked the Me'en area, killing thirty-four people, burning down 165 houses and taking away some cattle. The Me'en by that time were hampered by lack of weapons and ammunition. Government intervention or mediation failed.
12. A record kept of violent incidents with dead or wounded for the years 1990–2007 has at least one violent incident per month between either Suri, Dizi, Me'en, Nyangatom, Anywaa or highland villagers. I estimate the number of fatal casualties in these years to be at least 1,300, most of them Dizi. The total 2007 population of the groups combined is about 190,000.
13. The Derg government confiscated all their arms and the EPRDF regime only allows some militia to carry (registered) weapons. These are also of a lesser quality or older types. In contrast to the Suri, the Dizi are easily checked on the possession of arms.
14. An account on the plight of the Dizi, strongly written from their point of view and to be read cautiously, is the unpublished report of Addis Ababa University sociologist Abeje Berhanu, *The Dizi People and their Neighbours (a Report to the House of the Federation, Project on the Least Known Peoples of Ethiopia)*, Addis Ababa, 1999. See also Abbink (1993a).

drought the Suri went to the Dizi chiefs to plead and pray for rain, and a ceremonial sacrifice of a black ox was offered. When the Suri had a food shortage, cattle disease or other problems, they were permitted to enter the Dizi areas. Both groups still have economic exchange (cattle, pottery, iron products, grain, garden crops), but relations have steeply deteriorated in the last decade. For example, intermarriage has virtually stopped and the ritual alliance is no longer upheld. The Suri also see the Dizi as too closely allied to the state. Numerous efforts at mediation, including a few ceremonial reconciliation meetings – with a cattle sacrifice, cutting of the peritoneum and a joint meal – were tried in the past fifteen years, but none has lasted for long.

Suri and the state

Apart from the tension with neighbouring groups, Suri have serious problems with the state, which means in this context with local administrators and government agencies. This relationship has been tenuous for most of its history. Suri first met representatives of the Ethiopian state in the form of the imperial troops under *Ras* Wolde-Giorgis Aboyye in 1898, but did not enter into contact with them. The rule of Emperor Menilik II (1889–1913) condoned the autonomy of the Suri area, although Ethiopian northerners who came to settle in Maji and Jeba villages made excursions into Suri territory for hunting and trading and occasionally for raids. Under Emperor Haile Selassie, a cautious policy of rapprochement was started. Some administrative posts and primary schools were set up and, for a few decades, the Suri paid taxes. In terms of group relations, this period was relatively quiet, apart from the usual cattle raiding.

The Derg period is universally seen by local people as one of crisis and violence. The 1970s and 1980s, however, saw the emergence of wider regional tensions in northern Kenya and southern Sudan, as well as drought and food crises. In other words, certainly not all problems were due to government policies, although group relations between Suri, Nyangatom and Dizi deteriorated during the Derg era, and there was an increase in armed incidents. Government efforts to mediate did not succeed, and its drive to recruit local young men for the Derg army fighting in northern Ethiopia-Eritrea was deeply resented. Suri people were also victims of occasional violent reprisals from government forces, which deepened distrust towards the state.[15]

Under the EPRDF regime since 1991, efforts were made to mediate in local conflicts (see Abbink 2000c), in combination with an increased army presence in the area. The EPRDF forces in 1991 and 1992 took a cautious approach and tried to negotiate and create a dialogue with the people. They did not take punitive action after cases of Suri-Dizi violence (leading the Suri at first to qualify the soldiers as 'women', whom they should not be afraid of). But a series of incidents in 1993, in which Suri massively attacked Dizi settlements (in Kolu, Dami and Adikyaz) and,

---

15. One notorious incident was in 1986 when the Maji administrator invited a number of Suri men to a meeting in Maji to resolve a case of cattle raiding. When they were gathered, he had them tied up and shot. A number of them died, the others escaped. Suri took revenge a year later with attacks on people on the roads near Maji, whereby a number of people were killed.

more importantly, killed several EPRDF government soldiers in the Omo National Park and in Kibish, led to strong retaliatory action by the army. In a violent confrontation in late October 1993, an estimated 220 to 250 Suri, mainly women and children, died. Subsequently the violence diminished, but did not disappear. Suri antipathy and indifference towards the government (any kind of government) remained strong.

Since 1995 the Suri have had their own local council, and political co-optation and negotiation by the state continue with the gradual formation of a new group of young Suri leaders who have received jobs in the local and zonal administration. Even though they may be co-opted into a state structure where they have little real influence, Suri now do have some voice in the higher echelons of the state. They are formally represented in the local and regional administration and in the national parliament on the basis of an ethnic quota system. In this sense, their peripheral position as an 'ethnic minority' or 'nationality' has now become a kind of privilege, because other local people – for instance, the dispersed descendants of northern settlers in southern Ethiopia and living in the villages – are not politically represented.

Internal Responses to Violence within Suri Society

While the Suri have often inflicted untold suffering on their neighbours – like the massacres of Kärsi village in 1990 and Kolu in 1993, the gunning down of the young girls in 1993 in Adikyaz, the massacre in the Me'en area in early 2000 – the effects of such persistent violence bounce back and are felt within their own society. The dynamics of violence have an underestimated and unforeseen effect on the social structure of such small societies (see Hutchinson 1996, 2001; Elwert et al. 1999). Among the Suri, external crisis did not generate more internal solidarity against enemy groups; it seems to have had the opposite effect. First of all, the boundary between the two subgroups Tirmaga and Chai became more pronounced. Furthermore, Suri men fight among each other, over women, over cattle gained in a raid, over compensation payments, to settle scores at duelling places or while performing preliminary rituals for a raid. Often, excessive alcohol use (the local Suri beer *gèso* and *araqé*) is the trigger and, when automatic rifles are available, they are used. In all of such cases there are bereaved wives, children and parents, who mourn and demand redress (with the threat of vengeance killings always close). Empirical behavioural evidence, such as lower birth or child survival rates, family fragmentation, malnourishment, feelings of insecurity and grief, as well as effects like loss of labour power and loss of cattle due to more compensation payments to be made, reveal that the psychological toll of killing and death is high (see also Gray, Volume II). The frequency of grief inhibits people's long-term well-being and leads to what are called post-traumatic stress disorders. Married women especially have been increasingly voicing their concern and arguing with men over the new levels of violence.

The neighbouring Dizi, in their search for an explanation of Suri violence, often say that 'the Surma don't care about human life, it is nothing for them to lose someone'. But, as evident from often-heard private complaints, the Suri survivors do feel the loss and they do see the problem, but they don't know how to stop the violence. The young men have guns, carry them around everywhere and use them. A reconstructed image of the Suri male as an independent, assertive, gun-toting person afraid of no one and with the capability to use force to realize his aims has settled in Suri society. This new 'masculine' identity, which reinforces gender oppositions and conflicts, is a far cry from the former Suri 'warrior' persona of the old days as someone who defended the herds, who respected a code of the proper use of violence and who did not kill women and children on raids. Suri elders and also the Chai *komoru*, who are aware of the tragedy unfolding among their people, disapprovingly cited to me a few years back the ugly incident of a mass killing by young herders of cattle captured by the Nyangatom. When the Suri saw they could not recover the animals taken by the Nyangatom, hoping to hit the retreating raiders between the animals, they machine-gunned a large number of the animals from a distance. This was an unprecedented event.

Violence as an almost daily occurrence, or the threat thereof, has imprinted itself on the minds of the young generation: children see their parents or relatives falling away and their households crippled by the loss of labour power, affection and social support. As Chisholm (1993) has pointed out, there is a strong negative impact of such life-history experiences. Some of the negative effects are visible in social life.[16] I briefly mention several domains.

The topic needs further research, but it can be said that, in the wake of the newly articulated ideals of individualized masculinity, kinship ties are getting weaker, in that people in trouble get support from their kin less easily. Also the pressure on the bridewealth system is increased: not only have guns entered the system as part of the deal, but also arguments about the exact division are more acute. There is a growing demand by 'wife-giver' clans to receive more and faster as well. Thus, affinal alliances are under pressure: husband-wife relations become more strained and a tendency for males to neglect spouses and children becomes visible. Women with children are forced to use staple grains (sorghum and maize) for brewing alcoholic *gèso* beer, often for sale, rather than for feeding their own family properly. The health and nutritional situation of Suri children is, as a consequence, often precarious.

Also clan and lineage relations are under strain. The customary requirement to pay compensation cattle in case of accidental or wilful homicide is increasingly contested. As a result, the threat of feuding between the group of the victim and that of the killer, which always used to be 'bought off' with the compensation payment, has correspondingly grown. There are many unresolved feuds today. Some of them even extend into the kin group of a ritual leader: one *komoru* called Wolezoghi (from the Muge'i clan of the Tirmaga) was killed more than a decade ago, but the case – in itself unprecedented in that a *komoru* was killed by a fellow Suri – is still not resolved. Agnates of both perpetrator and victim regularly kill a member of the opposing clan.

The crisis in the age-grade system, specifically that of the loss of authority of the reigning age grade of elders over the youngsters, was evidence of internal disarray (see Abbink 1994, 1998a). The core relationship between the grade of the unmarried

---

16. For a revealing analysis of similar dramatic changes in Nuer society, see Hutchinson (1996, 2001).

young men, the so-called 'warriors' (or *tègay*, i.e. uninitiated), and that of the junior elders, called *rórà*, was disturbed because of the younger grade getting weapons and venturing out on their own. Since about 1989, Suri males have obtained automatic rifles (M-16s, AK-47s, GM3s and others). This has had a big social impact (see Mirzeler and Young 2000 on the Karimojong). The youngsters could ward off claims of the elders, and even stall their own initiation, because the new age-grade identity would have meant a responsibility they did not yet want. They extended their period of youthful exploits, of which using a gun was a major one.

Ceremonial duelling events have become more aggressive and dangerous, as the duelling grounds are occasionally arenas for shoot-outs that have nothing to do with the duelling itself (see Abbink 1999, 2000a). Duelling is a strictly regulated male combat sport among Suri of different herding units.

The position of the ritual leader or *komoru* is still there, but he is less heeded than in the past. Among the Tirmaga, *komoru* authority is slightest. The Chai *komoru* Dolleti V, who died in July 2001, was an authoritative person and had at least some restraining influence. He was replaced by a 'caretaker' *komoru* – one of his brothers – but an official successor is not yet installed.

In addition, an interesting cultural phenomenon is that the force and relevance of Suri ritual in itself are contested. While Suri society is highly ritualized where all significant social relations and statuses are established by ritual acts, in recent years their force has been waning. This was most readily apparent in the decade-long delay of the initiation ceremony for the senior age grade (see Abbink 1998a: 341–42). Also the cleansing ritual for homicide when an enemy group member is killed – and which requires the sacrifice of livestock – is less strictly performed.

Among the Suri today there is a pervasive sense of disruption and disintegration created by internal violence. It is deeply regretted, especially by elder Suri, and is seen as more problematic than the external violence. Perhaps people now realize fully that their society had always been ridden by latent tension and strife between kin groups and individuals. The system worked when the authority and role of the leading age grade and the *komoru* were still respected. But, due to dramatic conflict between Suri and the neighbouring pastoralists (especially Nyangatom and Toposa), the new wealth through the gold trade and the unforeseen aggressive power of youngsters of the *tègay* grade, it crumbled. The aforementioned killing in 2002 of their MP Guldu Tsedeke was another, and most acutely felt, illustration of this crisis.

Tourism and Missions

The story so far has made it clear that the Suri are not an isolated population but had contacts (though limited) within the broader regional setting and with Ethiopian society. At present they are caught up in an inexorable process of incorporation into a wider field of forces constituted by: (1) the regional-international conflicts in northern Kenya (Turkana-Nyangatom-state conflicts), leading to ethnic migrations to the north; and (2) the ongoing Sudanese civil war, leading to a flow of arms and ammunition and also to population movements; and (3) tourism and missionary influence. Here, I comment briefly on the latter. Together with the Mursi, the Suri are sought out by mainly Western tourists as one of the few 'really primitive tribes' still to be found in the African countryside – they are advertised as such in certain travel agency brochures. The encounter with tourists is aggressive and abrasive: in the absence of a shared language or a mutual interest in each other's humanity and because the principle of reciprocity is flouted during these encounters, the meetings are abrasive experiences for both parties (see Abbink 2000b). The Suri do not really accept the tourists, and they try to exploit them. They have not developed a relationship of simple commercial exchange (of pictures, goods and cash), but reproduce, time and again, an emotionally charged confrontation where the Suri express their contempt for the foreign 'other'. The tourists in their turn want to see their 'authentic, primitive tribe' but rarely want to be drawn further into their way of life, except in the form of takeaway photographs. This situation of 'double refusal' is rare in Africa (see Abbink 2000b).

Since 1994 there has been a mission station in the small village of Tulgi, in the Tirmaga-Suri area, staffed by American and Ethiopian missionaries. They have a programme of education, agricultural instruction, infrastructure development, medical care and Bible translation. There is a local church with regular services. The missionaries teach by example, and in the past years a community of about 120 Suri converts has emerged. It is too early to say what the impact of the Christian mission will be on the Suri. A previous missionary effort in the 1960s evaporated in the 1980s under the Derg regime and has left little legacy. But in the field of Suri local leadership and cultural orientation there may now be significant changes in the making. Suri are also reflecting upon how the Christian belief will affect their traditional notions and rituals, to which they remain attached. However, the need for change is felt, and the attraction of the evangelical message of peace and reconciliation, as well as its promise of social and economic connection to a wider global community of believers, is recognized.

A Non-adaptive System?

The Suri have a crisis between generations, an excess of small arms and a lack of internal peace between clans and lineages, and they live in permanent insecurity with regard to raiding threats, rainfall and food supply. They did not succeed in forging minimal alliances with neighbouring groups who share the natural and cultural space with them (such as cattle pasture, waterholes, forest items, game, cultivation sites and the old ritual burial and initiation places). Neither were they successful in getting the new regional government to take their problems seriously or to secure adequate representation on the level of the regional state (in Awasa). This is now improving, but the number of Amharic-speaking[17] or educated Suri is still very small. No doubt their being largely monolingual has inhibited communication with other groups, villagers and state agents.

So Suri society and economy are 'beleaguered': on all sides they are under pressure. In the south and west, the Nyangatom and Toposa are encroaching steadily and limit their pasture area. Most Suri herds (of both Chai and Tirmaga) have moved up north

---

17. Amharic is the working language of the Southern Regional State.

into the less favourable western Akobo Valley and near the eastern flanks of the Boma plateau. In the east they are confined by the Dizi mountains and government pressure, in the north by the Me'en (Tishana), who will not allow them to move in. While the Suri make raiding incursions into both areas, territorial expansion, settlement or herding there is not an option. In this respect, Suri are in a very difficult situation.

Their response to crisis has been one of disengagement, militant self-defence and violent entrepreneurship (the youngsters). Suri fear the government threatening their cattle wealth and will not submit to rules and regulations that impair the state and growth of their livestock or their autonomy. It is also a fight for respect to be shown to their identity and way of life as they are now, to which they, as one of very few peoples in Ethiopia, were able to remain attached. Their remote and inhospitable environment has so far helped them to keep up this attitude, but as this natural space is shrinking they will be forced to consider other options.

In the past the various groups in the Maji region were loosely allied and had open boundaries. Groups coexisted in simultaneous relations of violent incidents (raiding) and peaceful exchange of goods. Persons had bond partnerships and could be adopted as members in other groups. For example, some Tirmaga-Suri clans have Dizi and Me'en ancestry. The existence of shared rituals and codes of violent performance about which informants talk when referring to the 'pre-Kalashnikov era' before the early 1980s suggests that there was a kind of regional 'ethno-system' that in some way prevented excesses of violence between groups (although this is not to suggest that Suri society was a harmonious, integrated society). Alongside the cattle raiding, homicide and armed clashes that occurred, there were accepted ways of resolving and compensating for them. This system is known from other areas in the Ethiopian south. At present, such a system – which, however, should not be idealized either – is lacking or at least in steep decline, and an alternative adaptation based on indigenous principles but geared to modern conditions is not yet in sight.

In addition, the Ethiopian state has not succeeded in replacing the previous system with one of equitable dispute resolution or legal redress. Its policies are highly prescriptive, are imposed and rely on the ultimate use of force. As we know, the post-1991 regime started a policy intended to empower local 'ethnic groups' by engaging their representatives in local-level government, self-administration and education, although state agents retain control and follow their own agenda of political co-optation. The state is working through (self-created) loyal ethnic elites connected to the regional and national level, and is not aiming at grass-roots decision-making. The Surma Council is thus also – predictably – used as a conduit for implementing national policy (in the same way that the previous regime used the districts and peasant associations).

Conclusions: Changing Alliances?

The question as to the future fate of the Suri is a relevant one. If they do not succeed in the rehabilitation of their social relations and in restoring their measure of unity as a group under new local leadership, they will have a hard time surviving as a pastoral economy and as a people. Their economic base and their autonomy, not to speak of their group pride or identity, will come under great pressure, and they will remain a pawn in other people's games, including more powerful neighbours and, of course, the state authorities.

To summarize, the forces of change that reshape their society at present are: (1) persistent tensions and violent exchanges with neighbouring groups; (2) the Protestant mission, introducing the globalizing narrative of Christianity and trans-ethnic solidarity; (3) the slowly expanding system of state political surveillance and local/zonal administration and of education (primary schools); and (4) material and infrastructural developments: the need for periodic food assistance and disease control, trade needs, monetization and the opening up of the area by new roads. The simple fact of a new access road to the Suri area will have major consequences. A major all-weather road was completed from the town of Mizan Täfäri to Tum, the *woreda* capital of Surma and Dizi, with extensions planned further south to the Nyangatom area. This will do more than anything to 'open up' further the Suri area to external influences – state representatives, traders, missionaries, 'development' people and tourists – and will include the arrival of AIDS.[18]

So the Suri will not be left alone. Their area has a strategic position along an eighty-kilometre stretch of the Ethiopian-Sudanese border, and a growing economic relevance because of the gold trade, game resources and the potential of tourism in the two national parks (Omo and Mago Parks, developed with a major EU-sponsored initiative in the past years). While there is no doubt that Suri society and identity will be vulnerable to the hegemonic projects of the various categories of stakeholders, they will be forced to redefine their role in the region and to survive as agro-pastoralists through the forging of partnerships and alliances with other groups in the Maji region. This area remains a vast realm of several ethnic/social groups that are mutually dependent.

There is no doubt that these various hegemonic processes will reproduce the images of backwardness and 'primitiveness' about Suri among the various parties, most notably tourists, developers and the state. A cultural boundary of difference, if not condescension, will be confirmed. A significant indication here was the prohibition by the Ethiopian authorities of Suri ceremonial duelling a few years ago because it was 'too violent'. For the Suri, such a prohibition – which was not heeded – is a humiliating gesture. It foreshadows a complex, evaluative debate about 'good and bad culture', and also about the right to continue valued traditions that define people's identity vis-à-vis others. Due to the 'collectivization' of identity and of political rights in Ethiopia, these issues will be even more contested in ethnic terms. The ethnic administration model in Ethiopia, both rhetorically and practically, is thus crucial in reshaping perceptions of the importance of culture difference, redefining group relations and creating new forms of collective self-consciousness, whether these are based on the 'facts' or not.

To a large extent, Suri chances lie in exploring and building alliances and shared concerns with other groups in the Maji area: Dizi, Mursi and Me'en foremost, as they had alliances in the past and are interdependent today. Such a block of more than

---

18. Once this disease enters the ethnic communities in the Maji area, disaster looms for them because it will probably spread rapidly.

140,000 people might try to develop joint efforts around local problems of resource sharing and economic activity, hold periodic consultations, start projects of phased disarmament and enhance mutual cultural exchange (e.g. joining each other's collective rituals, which in fact was regularly done in the past). This is a daunting task ahead in view of the current tensions and distrust, but only then will these groups be meaningful vis-à-vis other regional players. The Ethiopian state should play a backstage role: not impose its model of 'development' and authoritarian political power, but only facilitate and work towards partnerships on the basis of equity, investments and a working legal structure. While such an approach would be in line with its officially declared policy of 'ethnic rights and autonomy', it is likely that the state will continue interfering and prescribing governance, 'peace' and 'development' in its own characteristic way. However, on the level of the newly emerging local elites (of Suri, Dizi, Me'en, Nyangatom, etc.), co-opted by the federal state, such new understandings of shared interests and cooperation may eventually be built, perhaps with the help of the emerging 'trans-ethnic' Christian community. Among the various ethnic groups, innovative leadership and the restoration of channels of peaceful negotiations and reconciliation rituals are necessary for it to happen: this means looking for continuities amidst change, and there are signs that this is indeed tried.

Currently, the small-scale societies in south-west Ethiopia are in a process of transition from ritualized, kinship-ordered social structures to territorial ones, partly defined by economic processes and by state bureaucratic discourse. This transition will lead to fragmentation and dispersal and to horizontal realignments among social strata of various ethnic groups and to normative pressure on their cultural traditions, thus subverting the model of ethno-cultural identity and organization that was proclaimed to be the future of the ethno-federal state. Trans-group alliances between group elites, based on concrete common interests and compromise politics and in tune with the cultural commitments of the ethnic groups' rank and file, will be a way forward. But this chapter suggests that social and political conditions at present are not conducive for these kinds of alliances to happen soon. As the killing of Guldu Tsedeke has shown, some of these small-scale societies in this area are not even at peace with themselves and have to solve their own specific problems of leadership and economic adaptation in order to establish such alliances.

Acknowledgements

I am deeply grateful to my friends and informants in the Maji area, Ethiopia, for their cooperation during fieldwork stays among them (1991–99). I thank various participants at the conference on 'Changing Identifications and Alliances in East Africa' (18–20 March 2002) at the Max Planck Institute for Social Anthropology for their questions and remarks on a first version of this chapter. Remaining flaws are mine.

Chapter 2

Resistance and Bravery: On Social Meanings of Guns in South-West Ethiopia

Ken Masuda

Guns as Socially Constructed Things

Guns are not only material and industrial products, they are also symbols of violence and triggers of memories, and sometimes they even form a historical index that reflects the political history of a particular location. While I was living among the Banna, I sometimes tried to touch or take a picture of a person's gun, but I was rarely successful because of the political and social attributes of guns.

The political attributes of guns are constructed in particular settings. One such setting is the Ethiopian centre/periphery situation, which stratifies the relationship between the centre of government and peripheral society, culturally, politically and economically. No one denies that anthropological studies in Ethiopia have been influenced by the so-called 'centre/periphery model', where, as Donham says, 'what was "peripheral" was always relative to a particular level of the hierarchy of centres' (Donham 1986: 24). On the issue of guns, the Muguji (Koegu) society was defined as relatively peripheral, as they bought out-of-date guns from the Banna, who are also peripheral to the centre of Ethiopia (Matsuda 1997). In addition, Ethiopia is peripheral to the centre in the modern world system.

The fact that all the guns described in this chapter were produced in Europe, the USA, China and North Korea reflects Ethiopian foreign policy and the history of the global political situation. However, the people who used those guns at each level of the world system are not necessarily conscious of their position in the whole. For example, the Banna collaborated with the British army and with Ethiopian soldier-settlers in an anti-Italian resistance. Those soldier-settlers had been previously recognized as oppressive 'enemies'. It does not seem to be that the Banna saw joining the Allied Forces as a case of switching 'from foe to friend', but that they saw themselves as fighting 'direct' enemies who threatened their lives at that particular time. An exploration of the history of guns and resistance among the Banna can lead to a better understanding of the way the Banna have identified themselves in their particular historical, economic, cultural and political settings.[1]

---

1. The following anthropological studies contain information on guns and societies: Pankhurst (1962, 1968, 1990); Abbink (1993b) on the Suri; Tornay (1993) on the Nyangatom; Turton (1993) on the Mursi; Matsuda (1997) on the Koegu; Chapple (1998).

This approach requires an examination of the cultural biography of guns. According to Kopytoff, the status of 'things' is always moving from being a commodity (with use value and exchange value) in a particular context into a new context where it is 'resocialised and rehumanised' (Kopytoff 1986: 65) and given a new identity and value (singularization). Kopytoff illustrates the process with the case of slavery: when an individual is a commodity, he becomes a slave, who is exchangeable. When the slave is acquired, he is reinserted into a particular community, where he takes on new identities and meanings and moves away from the status of a commodity. The potential is always there, however, for the slave to become a commodity again through resale (ibid.). This argument can be applied to guns: guns, produced in Europe and brought to Africa by traders, are exchangeable due to their value as commodities. However, no sooner are guns owned and used than they are immediately singularized and given new identities. The biography of a particular gun seems ambiguous, and is usually erased in the moving process by traders, governors and owners.

Meanwhile, the brand names and models of guns are connected to the memory of eras that would be traceable by mapping their names at points along a chronological table (see Table 2.1). In Banna, the periods of time, from Amhara domination to the post-socialist regime, are partially defined by reference to guns' names, memories of warfare and the types of guns that were used in them.

The Banna used guns in their resistance against the Ethiopian government and because of this the guns have political meanings. Through their resistance to the government the Banna expressed their collective self-image and through resistance they developed what they wanted to show and the way they wanted to be seen. Part of Banna identification, therefore, was accomplished through warfare and conflict. Important in this process were the use of and meanings attached to guns as socially constructed things.

The context in which the Banna identified themselves must be understood at two scales: the Banna and their neighbouring groups; and the Banna and the state (Ethiopia). Scholars' approaches to the former have been dominated by the ethno-system approach[2] and to the latter the centre/periphery approach. The study of guns is a case that enables these two perspectives to be combined.

Guns among the Banna: a Historical Overview

The Banna are an Omotic-speaking group living in the mountainous area between the Rivers Omo and Woito in southern Ethiopia (see Map 2.1). The population is estimated to be around 20,000.[3] According to Fukui (1984: 476–77), an ethnic category of Hamari, including both the Banna and the Hamar, is recognized by the

---

2. Fukui (1988) points out that the process of ethnic grouping can be considered to be an autonomous system of identification between groups, and proposes a concept of an ethno-system.
3. According to the Central Statistical Authority (1996), the estimated population of people speaking Hamerigna (the Hamar and the Banna) was estimated at around 40,000, but the exact number is unknown.

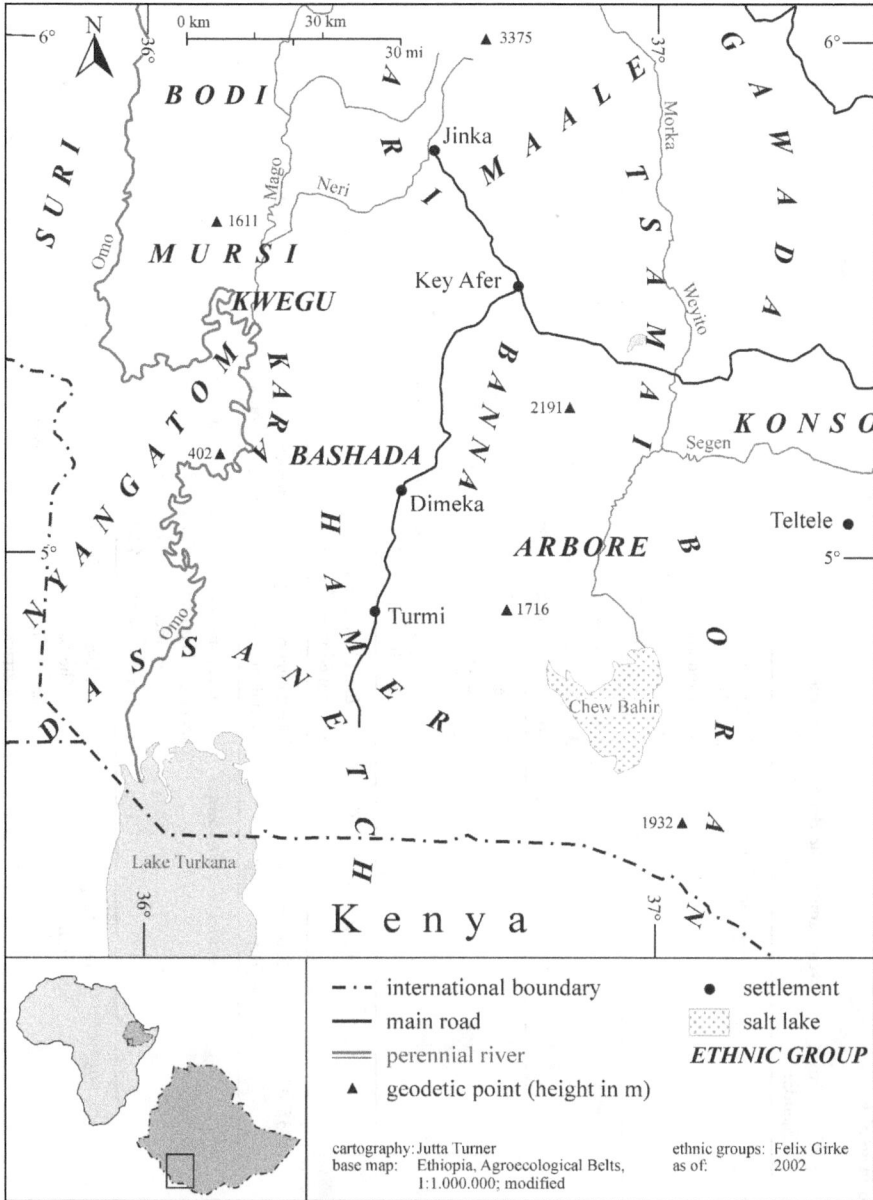

Map 2.1 Location of the Banna in south-western Ethiopia.

Bodi as an 'eternal enemy', and there is no way of resolution nor are there any rules for fighting with these groups. From the Banna perspective, the Bodi and the Mursi are the groups with whom relationships are hostile. The Dassanetch, Nyangatom, Arbore, Tsamay and Maale are also said to be enemies. In some villages in the lowlands, the Banna and Tsamay live together and practise intermarriage (Melesse

Table 2.1 A chronology of war memories

Date	Political Events	Gregorian Calendar	Events among the Banna	Comments	Names of Guns Referred to at the Time
Before 1890s	Before the Amhara domination	?	Korre banki (Korre war)	A mysterious group 'Korre' came from the east to attack and burn Banna villages. The Korre were said to be armed with spears and guns and to ride on horses. The Banna had no guns then. People ran away to the bush or went far away. The *Bita* of the eastern Banna, Doyo, died around that time.	
c. 1896–1935	The Amhara domination	C. 1896	Conquered	The situation of the early phase of contact is not clear. A *ketema* (garrison town) was established in Bako in the Ari area, and the government forced the Banna to pay taxes.	Naasi Massarya, Washtra
		?	Slavery	The people who could not afford to pay tax were commanded to present their children instead. Those children were later transferred to a slave market. A Banna, named Truga, was engaged in the slavery transaction and became rich.	Washtra, Laban
		?	Arrest of *Bita* Dore Wangi	The *Bita* of western Banna, Dore Wangi, was arrested and imprisoned in Addis Ababa.	
1935–41	Italian occupation	1935–41	Collaboration with Ethiopian *arbagnotch*	The Banna collaborated with *arbagna* patriots and resisted the Italian army and its Ethiopian collaborator, the *banda*.	Washtra, Kolle, Minishir, Alben, Dubai
			Atrocities of Italian army	The Italian army had only one Italian and others were *banda* (or Hamasien) soldiers, who were said to come from Tigray. Reportedly the *banda* raped Banna women, forced the people to work, and killed.	
1941–74	Haile Selassie period	1950s–60s	Resistance against the government	Gun confiscation by the government happened many times. The Banna killed governmental officers, police and merchants. The ambush gradually became a common style of attack.	Minishir, Alben, Dimotopor, Washtra, Gawutamura, Laban, Damba Mawzar, Dubai

Date	Political Events	Gregorian Calendar	Events among the Banna	Comments	Names of Guns Referred to at the Time
		Around 1960s	Cattle raiding against the Bodi	After obtaining a large number of guns, the Banna raided the Bodi for cattle. The raidings were followed by massacres.	
		Around 1960s	The murder of *Bita* Bura Dore	The *Bita* of the western Banna, Bura Dore, was indirectly killed by a Bako governor.	
		Around 1960s	Air raids	The Ethiopian air force bombed some places around the Banna-Hamar border.	
		1941–74	Governmental system	The government gave noble titles to Banna *Bitas*. *Bitas* were expected to act as local governmental chiefs.	
1974–91	The Derg Regime	1974–75	Arrival of revolution	Although neighbouring groups like Maale or Ari were forced to change, the Banna avoided strong socialist interference. Most people remember the age as a better time. Some Banna worked as *wereda* chief officers.	Dubai, Dimotopor, Chi'cha, Klash
		1974–80s	Conscription	Many Banna youths were conscripted and sent to the battlefield as Ethiopian soldiers.	
		1974–late 1980s	Infrequent guerrilla activities	Some groups of men engaged in sporadic guerrilla activities, including attacking the cars of government workers or merchants.	
1991–	The EPRDF Regime	1994–	New policy	After the promulgation of the new Constitution in 1994, each ethnic group was the given the right to autonomy. The position of the zone president in Jinka has been occupied by an individual from one of the local groups, the Ari, Nyangatom or Hamar.	Klash, Chi'cha, Dubai
		1992	The death of *Bita* Bezabih Bura	The *Bita* of the western Banna, Bezabih Bura, was shot by an EPRDF member with AK.	

1995: 123). Maale-Banna relations were very bad until the 1960s, but nowadays there is reportedly also intermarriage between them.

Social Meanings of Guns

The social implications of guns (*mura*)[4] should not be underestimated. Nowadays guns are used in bridewealth payments, as a supplement for cattle, goats, honey and money. Guns are connected to the construction of masculinity, and boys begin carrying a gun on their shoulder from the age of about fifteen. Male Banna youths are required to shoot wild animals around the Mago plain: through successful hunting, especially that of buffalo, lion and other big animals, a youth can prove himself to be brave and masculine. Guns are also crucial for cattle raiding and the killing of enemies. The killing experience is another opportunity for a man to increase his reputation. The values attached to such killings can be seen in some of the elders who, now over seventy years old, have patterned scars on their arms to show they have killed, and they also have their own 'killer's name'. The sound of gunfire also signals that someone has died, and it is common at a funeral to see condolence callers firing their guns at the sky. In addition, although I have only a few recorded cases, the gun can be used as a means of suicide in Banna society.

The guns that are carried by men are frequently empty of bullets. Most of the guns are decorated with a colourful piece of animal skin, and are of symbolic importance and value. Youths are also keen to buy and wear a *kansh* belt that has many pockets in which to keep bullets.[5] They are constantly interested in the maintenance and beautification of guns: some apply varnish or butter to its wooden stock. Bullets (*ushki*) are used as a substitute for money, and empty cartridges (*koiDe*) are used as decorations.

People like talking about guns. Whenever they have a conversation about hunting and killing, they always mention the name of the gun that was used. During dancing men gesture as if aiming at a target, and sometimes they actually fire guns. At the dance gatherings (*warsa*) held at midnight, the boys carry guns as ornaments on their shoulders.

The relationship between an owner and his gun can be likened to that between a husband and wife: when the Ethiopian government confiscated Banna guns and later sold them back during the Haile Selassie period, each Banna individual could find his own gun in a large pile. When someone found a gun belonging to his friend, he purchased it and returned it to him. The payment was subsequently reimbursed. No one could purchase and keep another person's gun, as owning another person's gun was considered equivalent to marrying another man's wife.

---

4. With this term, I exclude pistols (*shungcho*) from the discussion because I did not personally see any. I only saw the bullets for them.
5. There are two kinds of belt; a *kansh* has pockets that can hold several bullets, while a *kansh zinnare* is a belt with smaller pockets into each of which one bullet is inserted.

Gun Classification

Before describing the kinds of guns in Banna and their classifications, I shall first discuss the limitations of the investigation. The first difficulty that I encountered was that I had no opportunity to collect direct information concerning the older guns. These were mainly obsolete and were described in anecdotes of warfare, conflict and successful hunting. Through these discussions I learnt the names, the prices and the features that were recognized by the people. The names of the products, the manufacturing places, makers and specifications like calibre and bullet type could not be identified. In contrast to my own interest in gun type and places of production, I found that the Banna people paid little attention to these kinds of information; their gun cognition consisted of its form, its performance, the type of bullet and memories of warfare associated with it. The most reliable clue to identifying guns of pre-Kalashnikov generations is the names used by the people in the Banna or Amharic language. Some names are directly transferred from the original product names into Banna/Amharic (see Table 2.2).

It is possible to look in more detail at gun classification through the example of the Kalashnikov assault rifle (AK).[6] The most favoured AK among the Banna is the Chinese-made *Klash NatriBoqo*. It is distinguishable from the Chinese-made *Klash Dimpr* (*Nypr*), as the *Klash NatriBoqo* has a bayonet under the barrel and the *Klash Dimpr* has none. According to many, it is the bayonet that is considered attractive and makes the gun so popular. Differences in appearance do not always correspond to differences in the internal mechanisms and detailed specifications. For example, the category of AK called *Klash NatriBoqo* includes the Type-56 (AK-47) and Type-56 Press-Frame (AKM), both of which are made in China. These are not different in appearance but only in the manufacturing process and performance.

The Vicissitude of Guns

For the Banna, all guns are considered to be foreign things and they are imported to the area through various channels. Guns are classified locally into four generations. The first period (1890s–1930s) is when the Banna encountered the Ethiopian Empire, and the guns *Naasi-massarya* and *Washtra* were distributed. The next generation of guns, *Dubai, Laban, Damba Mawzar, Minishir* and *Alben*, were common in the second period, from the 1930s to the 1950s. This period was during and after the Italian occupation (1936–41) and these guns were brought by the Italian army. In the third period (1950s–80s), the only new introduction was the *Dimotopor*, but the older types coexisted with it. In the fourth period, from the 1980s, guns of the Eastern bloc like the *Chi'cha* (SKS) and *Klash* (AK) were introduced, during the time of the Derg regime and afterwards.

---

6. Developed by Mikhail Timofeyevich Kalashnikov, the first AK-47 (Avtomat Kalashnikova obrazet 47), using a 7.62 mm × 39 mm round bullet, was released in 1947. I generally use the abbreviated designation of AK or AKs. AKs have been produced in countries most of which belonged to the Eastern bloc during the cold war. All AKs I saw in southern Ethiopia were Soviet as were its successors, AKMs, and their variations. A comparatively new generation of AK (AK-74), which uses a new type of bullet (5.43 mm × 39 mm), had not been discovered.

Table 2.2 Guns that have been used among the Banna

	Name in Banna	Name of Product, Manufacturing Country	Calibre	Age	Remarks	References
1	Naasi Massarya	?		1890s–?	This gun is said to be the first the Banna encountered and the one the Ethiopian army used when it came to the area for the first time. It was a single shot gun. *Naasi* means child in Banna, and *Massarya* means gun in Amharic.	
2	Kongo	Russian?		1890s	Although its performance had a bad reputation, many Kongo were distributed. The origin of the name is unclear.	Matsuda (1997) says Kongo was a Russian musket. Kurimoto (1992: 13) reports that Russian guns named *moscob* were used among the Anywaa.
3	Washtra, Astra, Wajigra	French Gras?		1890s	It is quite difficult to identify what the Washtra (or Astra, or Wajigra) was. It possibly existed before the Italian invasion, or was brought by Italians. It was also a single shot gun or five-shooter. I think there were two kinds of Washtra: one was long and black and called Kolle; the other was short and metallic and called Gawumura or Gawutamura because its colour resembled that of a *gawu* brass bracelet. The Washtra was also called Orgo, and the term *orgo* means short in Banna. It is possible that the Washtra Orgo was the same as Gawutamura. It is said that the first man who shot an elephant used an Orgo gun.	The local name Washtra, or Wajigra, resembles the Abu Gigra gun of the Nuer (Hutchinson 1996: 11) but I have no evidence to confirm a connection. Kurimoto (1992: 13) also mentions a Wajigira rifle among the Anywaa and says that the name originated in Fusil Gras. According to Pankhurst (1968: 600), Ethiopia was provided with many French Gras from 1895, and the guns were used in the Battle of Adwa in 1896. Fusil Gras was produced from 1871, and Pankhurst (1962: 172) suggested that Fusil Gras was called *wajigra* in Amharic around 1990. Alexander Naty (1992: 76) mentions that the Ethiopian army was armed with Wajigra Moscob when it conquered the Ari of southern Ethiopia.

Name in Banna	Name of Product, Manufacturing Country	Calibre	Age	Remarks	References
4 Dubai	Czech or Belgian		late 1930s		According to Pankhurst (1962), Ethiopia purchased many guns from Belgium, Czechoslovakia, Yugoslavia and Japan in order to defend themselves from attack by Italy. The Dubai appears similar to the British 0.303-inch Short Magazine Lee-Enfield, which was used during the First World War.
5 Laban	French Lebel?		1930s?–1960s	The Laban was a six-shooter and its barrel was long. The Banna began to use Laban during the Italian regime, but, because of its price, only rich men could afford to buy it. Ordinary people used Washtra or Gawutamura at that time.	Because of the resemblance of pronunciation, the Laban could be the French Lebel, produced from 1886.
6 Damba Mawzar	German Mauser?		1930s	The Damba Mawzar probably came to the Banna in the same period as the Laban, but it was rarely mentioned by informants. It may be a Mauser gun from Germany.	The German Mauser was one of the most famous firearms used during both world wars; many were exported to Africa. In Ethiopia, the Mauser was used in the Battle of Adwa (Chapple 1998) and during the anti-Italy war in 1935 (Pankhurst 1962: 174).
7 Minishir	Austrian Mannlicher?		1930s	Minishir was brought by Italy and arrived in Banna later. Banna informants recollected that one Minishir was bought for fifteen cattle in the 1950s and 1960s.	Pankhurst (1962) says that a gun called Minishir had already been used in Ethiopia, but the product name was unclear.
8 Alben	Belgian Albini?		Late 1930s	The Alben was also brought by Italy. It was a six-shooter and powerful enough to shoot over a long distance. There might be a long and a short form.	

Table 2.2 *Continued*

Name in Banna	Name of Product, Manufacturing Country	Calibre	Age	Remarks	References
9	Dimotopor, Dimopor, Dimas	British Lee-Metford?		1950s–1980s	The Banna came into contact with the Dimotopor during the Haile Selassie regime. The heavyweight gun had two variations: one was the long Guncho Dimotopor, which was used by British army to fight Italy; the other was the short Afadist Dimotopor, which was formerly used by the Ethiopian army. Dimotopor was a major weapon in the 1950s to 1960s.	Matsuda (1997) reports that there was a gun called 'Dimmotfer' in the past and says it was a DM-4, which was one of the DM products made by the German Deutsche Metallpatronenfabrik in the 1890s. It is also possible that the Banna Dimotopor was a British Lee-Metford. Pankhurst (1962: 170) writes that Ethiopia imported many Lee-Metfords in the 1890s, and I suggest that some Lee-Metfords became Dimotopor.
10	Otomatik	US M-1 rifle			There were only a few Otomatik. The name is derived from the English automatic. It is thought to have been imported from Somalia. The Banna recognized the gun as that of the police.	The M-1 rifle was used by the US army during the Second World War.
11	Chi'cha	Russian SKS-45 and Chinese Type-56 Carbine	7.62 mm	1980s	At present, there are many Chi'chas. Most of them have a serial number in Chinese letters. Ten AK bullets can be charged at a time. According to a rumour, the first man who bought a Chi'cha paid for it with eighty head of livestock.	The meaning and origin of the name Chi'cha is vague. It is called SKS (*es kei es*) or Chi'ch in the Amharic-speaking world.
12	Princhi'cha	Chinese Type-63	7.62 mm	1980s	Few people own Princhi'cha. Its shape is similar to Chi'cha, but it can shoot twenty bullets.	

	Name in Banna	Name of Product, Manufacturing Country	Calibre	Age	Remarks	References
13	Klash	Made in the Eastern Bloc	7.62 mm	1980s	The term Klash indicates the Soviet-made AK-47, AKM and variations. Its banana-style magazine is distinctive. There has been a plentiful supply of bullets from southern Sudan, which cost three birr in 1999 (and five birr in 1993). AKs have been the main weapon of the Ethiopian defence force since the 1970s, when the country had strong ties to the Eastern bloc, and it became popular among the Banna after the collapse of the Derg. A Klash cost five cattle in the late 1980s, but, because of supply increase, people paid only two to five cattle or four hundred to one thousand birr in 1999.	AKs are called Klash among Amharic speakers too.
14	Klash Labana	Russian AK-47 and North Korean 58, etc.	7.62 mm	1980s	Klash Labana seems to be a Soviet AK-47. It is difficult to identify variations in detail.	
15	Klash NatriBoqo	Chinese Type-56 (AK-47) and Type 56 Press-frame (AKM)	7.62 mm	1980	This is the most popular copy model of AK-47 and AKM. Most of them have a bayonet.	Matsuda (1997) mentions that the Koegu had a kind of AK called Natolibok and identifies it as made in China or made in Yugoslavia. The name Natloboko is derived from a cattle hide pattern *natdobo* (personal information by a Nyangatom friend).
16	Klash Sholo	Russian AKM	7.62 mm	1980s	AKM was a revised model of AK-47. Its muzzle shape is distinct from that of AK-47.	
17	Klash Nimpr (Dimpr)	Chinese M22	7.62 mm	1980s	Chinese model of AKM. The Banna identify Nympr from NatriBoqo by the form of muzzle.	

Table 2.2 *Continued*

	Name in Banna	Name of Product, Manufacturing Country	Calibre	Age	Remarks	References
18	Klash Kuntsa	Hungarian AMD-63 or Romanian AIM?	7.62 mm	1980s		
19	Klash Chaagi	?	7.62 mm	1980s	The folk term *chaagi* is a basic colour term for green or blue, so it is possible that the Banna named Klash Chaagi for a kind of AK that may be a Hungarian AMD with a pearl blue stock. According to the people, Klash Chaagi had an additional grip under the barrel.	
20	Klash Chamma	Polish?	7.62 mm	1980s	This gun has a grip under the barrel.	
21	Klash Matris	PRK and variations	7.62 or 5.45 mm	1980s	AKM was modified to be an RPK machine gun with folding legs under the long barrel. RPK are used by the police and army; only a few are owned by Banna.	

In 1993, old guns like the *Dubai* were most common, and the *Klash* was still a luxury item. This seems to have changed around 1999, when most men had either a *Chi'cha* or a *Klash*, and only a few had a *Dubai*. In the late 1980s, a Banna man could visit Teltele, beyond the River Woito, and pay fifteen cattle for a brand new AK. A famous anecdote relates how the first *Chi'cha* owner paid eighty goats and sheep for it. By 1999, the price of these guns had dropped to two to five cattle for one gun, or in money, around one thousand birr.[7] The prices of bullets changed from five birr in the early 1990s, to four birr in 1993 and three birr in 1999, as supply increased from southern Sudan.[8]

Changing Alliances

In this section, I examine the Banna memories of conflict and warfare along with the chronological usage of guns. Chronology in this context does not mean an objective, scientific and authentic scale, measured with homogeneous ticks, but refers to varying, emic classifications and meanings. I have to justify the reasons for choosing such an 'unreliable' material of study. The first, somewhat negative, reason is the lack of written historical sources. The second, more positive, reason is that in studying Banna identification it is preferable to use oral histories, in which memories and interpretations are mixed (Masuda 1997). In the description below, I combine these histories with other information from bibliographic research.

Amhara conquest (c. 1890s–1936)

Banna memory concerning first contact with the Ethiopian Empire is ambiguous. Ras Wolde-Giorgis was the person who led the army advance to the south and, with the support of Dejazmatch Tessemma Nadew and Dejazmatch Damtew Ketema, began the conquest over the region. It is not surprising that the Banna do not know the name of Wolde-Giorgis, as control over the Banna was not strong in comparison with that of the Ari and the Maale. From the early phase of encounter with the northern regime, the Ari and Maale social systems were forced to undergo fundamental changes (Alexander Naty 1992; Donham 1994). A garrison town *ketema* was established in Bako, southern Ari, after the conquest, and thereafter the *ketema* functioned as a local centre of government. Small police substations were built throughout southern Ethiopia, even in the Banna-Hamar area. The stations not only raised an Ethiopian flag and distributed malaria medicine, but provided material images of 'modernity' (Donham 1999).

In the early phase of contact, when Ethiopian soldiers began to penetrate into Banna land, many people left the villages and ran away into the bush. Some of them were killed for their supposed commitment to resistance activity. The expression 'ran into the bush' (*qaunte gobidi*) is a typical phrase that I have heard repeatedly during conversations. After bloody battles, the Empire managed to appease the Banna *Bitas* (priest-chiefs) by giving them many gifts, including guns, and people began to pay

---

7. This was possibly triggered by the 1998–99 famine.
8. Bullet price in 2001 was five birr because of shortage of supply from southern Sudan.

taxes. The *Bitas* of western Banna had stronger connections to the government than the eastern *Bitas*. Amhara soldiers and officers sought natural resources like ivory, which had already disappeared elsewhere. After receiving guns such as the *Washtra*, the Banna, who had known only trap hunting until then, began shooting wild animals. This increase in hunting activity caused a decrease in wildlife in the south.

While Banna *Bitas* kept good relations with the government, dissatisfaction about tax payment was smouldering among the ordinary people, whose children were sometimes taken and sold if they could not pay taxes by a deadline. It was around this time that Banna resistance against Amhara soldiers and *neftenya* (soldier-settlers) took place, mainly by ambush. At this time guns were also usually hidden in the bush.

Italian colonization period (1936–41)

The actions of the Italian army are described by the Banna as brutal. The army had only one Italian commander; most of the soldiers were Ethiopian collaborators with Italy and were known as *banda*.[9] The Banna direct their resentment to the *banda*, not towards Italy, because of their memories of the cruel experiences that the *banda* forced upon them. Many people were compelled to carry things, to supply foods and so on. Some exhausted people were recognized as useless and were shot. The most infamous story of *banda* mercilessness is of rape: some *banda* soldiers raped a Banna woman in front of her husband. He was forced to watch and later was hanged from a tree.

Confronted with these miseries, the Banna were gradually absorbed into partisan activities. They allied themselves with Amhara patriots (*arbagnotch*), who had been *neftenya* settlers. Together they waylaid the *banda* soldiers in the bush. People who did not join the guerrilla activity moved to the south. Some of them passed across the Ethiopia-Kenya border and joined the British army (*jambo*), where they supported the effort to prevent the Italians from marching further south.

The Italian occupation of Ethiopia brought with it brand new and contemporary European guns (*Dubai, Damba Mawzar, Minishir, Alben*), which the Banna obtained from the soldiers they had killed. The 'military revolution' (Parker 1988) that came with the more rapid-firing guns spread into Ethiopia. In Banna, the increase in gun possession was also accompanied by a high frequency of cattle raiding.

Haile Selassie regime (1941–74)

Although Emperor Haile Selassie instituted the new constitution in 1931, his presence as Emperor became recognized by the Banna people only after the liberation of 1941. The relationship between the Banna and the Ethiopian government was very hostile during the period of rule that followed.

During this time, the Ethiopian government recognized *Bitas* as *balabbats*, and conferred some titles of Amhara aristocracy on them. The title of *Grazmatch* was given to authentic *Bitas*, and the title of *Balambaras* was given to acting *Bitas*, who were *Bitas*' younger brothers during the *Bita* vacancies (see Figure 2.1). Through this process, Banna *Bitas* were incorporated into the class of Amhara nobility.

---

9. The term *banda*, originally from Italian, signified Ethiopian collaborators (Bahru 1991: 174).

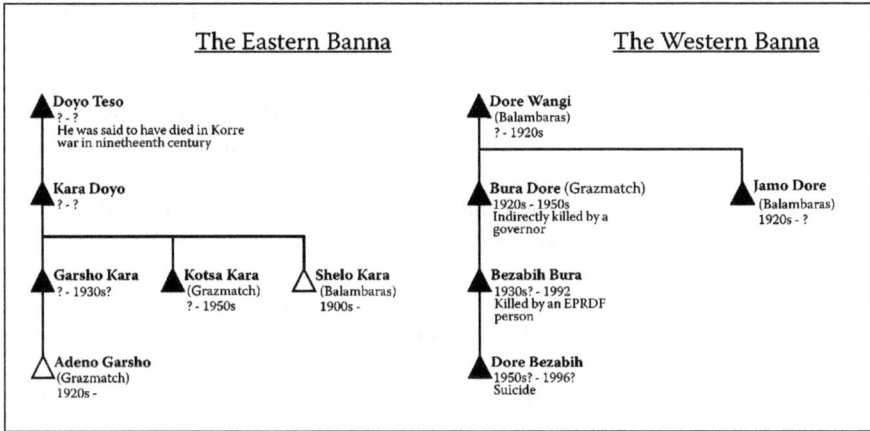

Figure 2.1 Genealogies of *Bitas* with their offices.

On the other hand, the Banna continued to resist the northern Ethiopian regime by attacking officers, policemen and, sometimes, traders. They referred to all these people as *gal* ('Amhara' or 'enemy' in the Banna language). People recollected that, 'when we saw shoe prints on the ground, we quickly recognized them as belonging to an Amhara because the Banna did not wear shoes then. Tracing the prints, we killed the men immediately after we found them.' In addition, 'when we heard the sound of a car, we immediately ran into the bush and watched. Whenever they acted suspiciously, we killed them' (interview in Bori, 1999).

During this time, 'everything', said a Banna, 'was required and taken by police. They were like thieves' (interview in Bori, 1998). Some Banna were appointed *Chika Shum*, an official position to collect taxes. They sided with the government and always accepted bribes. It is said that not only *Chika Shums* but also some *Bitas* were exploiting ordinary people. In brief, Banna society, in a sense, began to fracture politically.

The Banna clearly remember the cases of gun confiscation, which were organized by Amhara officers at this time under pretexts like 'war is forbidden by the law' or 'prevention of anti-governmental activity'. This case apparently happened in the 1950s: a lieutenant, residing in Bako, came to the Banna and hanged Bura Dore, the *Bita* of western Banna, and Kotsa Kara, acting *Bita* of eastern Banna, and then ordered all guns to be brought in. People complied with these orders. The guns were kept in storage for several months and then sold to their former owners. Around the time that similar gun confiscations were carried out, people, especially youngsters, became more involved in resistance with violence.[10] Violent resistance took place with the use of guns, but the resistance was also organized covertly and took more

---

10. It is interesting to compare this with the Karamoja case, where the spread of AK-47s after 1979 and the involvement of youngsters introduced a new dynamic, again favouring the emerging of warlords and the decline of the elders (Mirzeler and Young 2000: 419).

subtle forms: people kept silent to the police and hid murderers. The government called the Banna *shifta* (outlaws) and labelled them 'the people who were never governed'.

Bita Bura Dore was killed by the governor, Asfaw Gebre Amanuel, who was in position in 1957. Bura Dore's death was followed by his small brothers and sons firing guns at police and governmental officers as retaliation, saying 'we are not dogs'. After a series of small battles, the Ethiopian air force decided to bomb Banna and Hamar.[11] Planes bombed an area around the border between the Banna and the Hamar intensively, and the government sent its military forces in by land. The Banna fought these soldiers with the guns they had obtained during and after the Italian occupation. This battle brought Ethiopian soldiers' *Dimotopor*, which were recognized as being made in Britain.

It was during this period that inter-ethnic cattle raiding became even more frequent. The Banna raided the Mursi and the Bodi in the west and the Maale in the north. Inter-ethnic warfare is considered to have changed radically at this time, as the shift from spears to guns led to an increase in the number of victims of raiding. The government intervened in the conflicts, arrested criminals and forced them to return the cattle. The rural people, however, believed that the government or officers appropriated those cattle. There are similar kinds of government interventions and assertions reported in the Hamar case by Lydall and Strecker (1979a: 90).

The Derg regime (1974–91)

In Maale and the Ari society, the new Derg government acted severely and abolished the chiefs. In Banna, by comparison, the relations with the government were peaceful. The reason for this was that the Banna land system, which does not admit individual landownership, only use rights, did not conflict with socialism. Powerful newcomers, who had been resented by the Banna, like the *neftenya*, were abolished. Moreover, there were some opportunities for some educated southerners to become officers and administrators in the local government.

The most fundamental change occurred at this time, as Banna society became more incorporated into the state than ever before. While the area in northern Banna, formerly ruled by the *neftenya*, was returned to its original owners, those people were simultaneously reorganized into peasant associations, which functioned as the lowest level of the governmental system (*kebele*).

Ethiopia in the 1970s and the 1980s was always at war, and the government conscripted Banna youths through the peasant associations, trained them and sent them to the war front. Socialist Ethiopia was provided with a huge number of firearms from the Eastern bloc. By the end of the 1980s, SKSs (*Chi'cha*) and AKs (*Klash*) had arrived in Banna. Most of the guns used there were still old types like the *Dimotopor* or the *Dubai*, however.

---

11. Despite their recollections, I could not find any documented evidence of the bombing. As Gebru mentions, the Ethiopian air force often bombed to suppress anti-government riots in Gojjam in 1968 and Bale in 1969 (Gebru 1991: 148, 184).

Post-socialist regime (1991–)

The collapse of the Derg and the establishment of the EPRDF government led to new developments, not only in the centre but also in the southern periphery. The political power of southerners became much stronger because the new Constitution declared the self-governance of regions by local people. The administrator of the Banna area[12] is an educated Banna and the relations between the Banna and the government appear to be amicable.

The peace and the Banna's trust in the government are no more than superficial, however. After the news of the Ethiopia-Eritrea war (in May 1998), Banna youths did not go to market because they were wary of conscription. The Banna still conceal their guns from the police. They never bring guns to the market, or, if they do, they leave the guns outside the town. The possession of a gun by a Banna is given unspoken consent as long as they stay peaceful. Nevertheless, it is certain that the police will arrest someone who is discovered taking part in the gun trade. Conflict has been intermittent: the most strained situation occurred when Bezabih Bura, the *Bita* of western Banna, was shot in the town of Kako in 1992. The man who shot him was said to be from the EPRDF. Soon after Bezabih's death, his relatives attacked Kako and killed some policemen and teachers.

For a while after the collapse of the Derg regime, the supply of AKs became greater. This made it easy for the Banna to obtain guns peacefully. A large number of *Chi'cha* and *Klash* were brought and sold by traders, and the Banna lost many cattle as payment. Old types of guns like *Dimotopor* or *Dubai* were replaced by *Chi'cha* and *Klash* (see Table 2.3). From a military point of view, the Banna armaments developed almost equally to those of the police, but the latter were equipped with newer technologies like sub-machine guns and RPKs, known locally by the name of *Klash Matris*.

With respect to inter-ethnic relations at this time, hostile relationships seem to have disappeared. The Banna have stopped raiding the Maale for cattle, and recently there have been several cases of intermarriage between the two groups. Because of their remoteness, the Banna are rarely in contact with the Bodi and the Nyangatom. The only group they still have strong hostility towards is the Mursi and there have been some killings in the Mago Plain in recent years. Turton (1993: 167) reports that, during some five years from April 1985 to July 1990, twenty-seven Mursi were killed by Hamari (Banna and Hamar together).

Historicity of Identification

For attacking and raiding neighbouring groups, the Banna used two modes of attack, *sula* and *banki*. On the one hand, *sula* is a small-scale attack usually done by a small group; on the other hand, *banki*, which generally continues over a long period, indicates a large-scale attacking mode, executed through detailed operation plans and ritual procedures by military organization of the age-grade system. After warnings of Mursi attacks, the Banna began preparations for conflict by purchasing guns and

---

12. Banna-Kule *wereda* (district), then Hamar-Banna *wereda*, and then Banna-Tsamay *wereda*.

Table 2.3 Guns owned (1998–99)

No.	Age of Owner	Guns' Name in Banna	Type of Product	Remarks
1	40	Klash Sholo	AKM (USSR)	
2	30	Klash Sholo	AKM (USSR)	Purchased with four head of cattle
3	35	Klash Sholo	AKM (USSR)	
4	?	Klash NatriBoqo	AK (China, export model)	Serial number 386 56 3640874.
5	50	Klash NatriBoqo	AK (China, export model)	Purchased from a merchant with four head of cattle. Sold at 2,000 birr later.
6	45	Klash NatriBoqo	AK (China, export model)	The stock made of light-coloured wood. The muzzle and bayonet decorated with animal skin. Purchased from a brother at 1,700 birr.
7	25	Klash NatriBoqo	AK (China, export model)	
8	25	Klash NatriBoqo	AK (China, export model)	Decorated with skin of *murja* (sitatunga or kudu).
9	20	Klash NatriBoqo	AK (China, export model)	
10	20	Klash NatriBoqo	AK (China, export model)	
11	17	Klash NatriBoqo	AK (China, export model)	Serial number 386 56 3545654.
12	30	Kash Labana	AK (North Korea)	Purchased from a Banna with four head of cattle. Same appearance as Soviet AK-47. The muzzle decorated with skin of *murja.*
13	16	Klash ——	?	Local name unidentified.
14	?	Klash ——	?	Local name unidentified.
15	?	Klash ——	AK (China, domestic model)	Type 56 model with folding stock. The stock made of plastic. Local name unidentified.
16	55	Klash ——	M22 (Chinese AK-47)	Without bayonet. Local name unidentified.
17	45	Klash ——	Chinese Type 56-1 model?	With folding stock. Local name unidentified.

No.	Age of Owner	Guns' Name in Banna	Type of Product	Remarks
18	25	Klash ——	?	Local name unidentified.
19	20	Klash ——	AK-47 (Bulgaria)	The stock is made of plastic. Local name unidentified.
20	90	Dubai	?	Data in 1999.
21	40	Dubai	?	Data in 1998.
22	35	Dubai	?	Data in 1998.
23	50	Chi'cha	Chinese Type-56	
24	50	Chi'cha	Chinese Type-56	
25	50	Chi'cha	Chinese Type-56	Serial number 0136 五六式 25007567 in Chinese characters. The bayonet decorated with skin of *gash* (warthog).
26	45	Chi'cha	Chinese Type-56	The trigger decorated with skin of *guumi* (antelope). The bayonet decorated with *ukuso* (mongoose or civet).
27	40	Chi'cha	Chinese Type-56	
28	40	Chi'cha	Chinese Type-56	
29	40	Chi'cha	Chinese Type-56	Data in 1999.
30	25	Chi'cha	Chinese Type-56	Purchased from a Banna with three head of cattle.
31	18	Chi'cha	Chinese Type-56	Purchased from a merchant with four head of cattle. The muzzle and trigger decorated with goatskin. Number 15377 stamped on the stock. Serial number '五六式 24005607' with Chinese letters.
32	17	Chi'cha	Chinese Type-56	Skin covering from the muzzle to the body. The muzzle also decorated with animal hair.
33	50	Princhi'cha	Chinese Type-63	Imported through Mursi area.
34	50	Otomatik	M-1(USA)	

bullets, sending out scouts and driving cattle away. The 'army' consisted of young 'warriors' and old 'commanders' classified roughly by the age-grade system, while boys who were not recognized as fighters were engaged in guarding cattle. The Banna age-grade system is said to be imported from the Nyangatom and functioned well in the past. Nowadays, however, people are not attentive to the name of each grade or the order in which it comes. It seems possible that there is a correlation between the decline of the age-grade system and a recent decline in inter-ethnic conflicts.

Inter-ethnic conflict and relations were often interpreted previously as locally produced within a self-contained system (ethno-system: see Fukui 1988). From the ecological point of view, inter-ethnic relations and their history are interpreted as a result of pressure on natural resources. Without denying the ecological approach as one way to deepen the discussion, I would like to suggest the importance of a 'political ecology' approach. This would see inter-ethnic warfare not only as a local matter, but also as one that is linked to national and global processes. Here, it is connected also to Banna resistance to the state and to colonial power.

The example of the Mursi-Nyangatom conflict from 1987 to 1992 shows that the political conditions around the state influenced the peripheral situation. The Nyangatom attacked the Mursi in 1987 with AKs they had bought from southern Sudan. The groups had been hostile to each other before then, but this AK-armed attack inflicted serious damage on the Mursi. The Mursi became ready to take revenge on the Nyangatom in March 1992, when they obtained AKs that were taken from an Ethiopian military armoury after the collapse of the Derg regime, and after they had recruited new warriors at an age-set ritual in 1991. Being armed with AKs made it possible for the Mursi to attack the Nyangatom (Turton 1993).

These examples of the Banna, the Nyangatom and the Mursi suggest the following points: (1) the introduction of new types of gun changes existing inter-ethnic relationships; (2) the guns were from the civil war in southern Sudan and the Ethiopian armoury; and (3) the condition of armaments among each group reflects domestic political situations and foreign relationships. Moreover, it should be kept in mind that guns were (and are) a violent device of social control, exploitation and oppression, and that they are found in marginal areas, where they are used in inter-ethnic conflicts and in resistance to the state.

In addition, I should point to the possibility, already suggested by several authors, that inter-ethnic conflicts were triggered by pressure from outside. A Hamar elder, Aike Berinas, said that the decision to attack the Mursi was taken by his father, Berinas. The Banna tell stories indicating that it was the government that made Dassanetch/Hamar-Banna relations hostile, and that an Amhara officer, fuelling an already hostile relationship, led Banna warriors to raid the Maale.[13] Therefore, there is a situation where inter-ethnic relations result from the intermingling of the ethno-system with the outer influences of politics and materials such as guns.

---

13. Lydall and Strecker (1979b: 25–26) introduce an anecdote which shows that a Bako governor, Dedjazmatch Biru, forced the Hamar to attack the Mursi.

The fact that the times of inter-ethnic warfare from the 1950s to the 1970s overlapped with periods of anti-governmental resistance leads to three hypotheses: (1) although two kinds of warfare (with the state and with the neighbouring groups) were different both in context and fighting style, they were contemporaneous; (2) one of the factors increasing cattle raiding during the Haile Selassie period was the desire for capital to pay for guns; and (3) gun-purchasing possibly caused a decrease in numbers of Banna cattle.

The word *gal* (enemy) has two implications: it refers to hostile neighbouring groups like the Mursi, and also to Amhara people and northerners. Banna people usually say '*gal*' to indicate the Amhara category, which includes officers, policemen, military commanders, merchants, bar owners, teachers and so on. The Banna usually choose ambush as the mode of attack against *gal* Amhara. This type of activity could be called a guerrilla-style attack, which means physical violence by a sub-state subject. The Banna guerrilla-style activities did not have an ultimate purpose, for example the collapse of government, but remained as revenge for personal loss and exploitation.

According to Gebru Tareke (1991: 126), one of the distinctive features of resistance activity among Ethiopian peripheral societies is that, like the Bale resistance of south-eastern Ethiopia in the 1960s, their military activity was not successful in promoting any social changes because of a lack of leadership with long-term vision. In Bale, the Oromo-Somali obtained new guns and were militarized in both technical and organizational aspects with assistance from Somalia. In contrast, among the Banna, neither modern militarization nor the organized military operation based on the age-grade system functioned in resistance against the state. Although some Banna were engaged in the Ethiopian national army during the Derg period, the modern military system was not introduced into the Banna. In the following paragraph, I point out three reasons for this.

First, Banna soldiers were at the lowest level in the army, as a centre/periphery structure also functioned in the military. Secondly, the Banna's relations with their neighbours had become calmer by the time the Banna joined the Ethiopian army. Thirdly, the modern organizational system was so alien to the Banna that they could not adopt it. On the contrary, they adopted the age-grade system of their ideal enemy, the Nyangatom. The decision as to whether or not they adopt the enemy's military system depends on the degree of 'otherness'. While I recognize many differences between the Nyangatom age system and the Banna one, the Nyangatom was possibly recognized as being closer to themselves than the features of Amhara society. A Banna told me, 'We, the Banna, seek cattle and human life, but the Amhara seek land. We do not own land because all land belongs to *barjo* [good fortune or well-being]' (interview with *Bita* Adeno Garsho, 1999). The Banna and the Nyangatom share the same values of their cattle-based culture.[14] The narratives

---

14. In an article, Gray points out that the Karimojong, who once lost their identity as cattle herders, regain it through cattle raiding after firearms became available (Gray 2000).

point out the linkage between a lack of landownership with the idea of *barjo* to stress a cultural difference. Such recognized cultural differences were also reasons for the avoidance of the modern military system.

Guns as Historical Index

The hypotheses above require an investigation of why the Banna were enthusiastic about guns. However, this question immediately leads to circular reasoning: is it that increasing guns caused cattle raids and attacks, or that increasing cattle raids caused demand for and a flood of guns?

Guns replaced the spears (*banki*) and arrows (*om*) that were previously used for violence.[15] This replacement, however, did not undermine the Banna idea of the connection between violence and masculinity.[16] The construction of Banna masculinity is based on violence, as a man who cannot beat his wife will be laughed at (Lydall 1994).

It is interesting to compare the case of the Ari, reported by Alexander Naty, to the case of the Banna on gender and violence. Alexander Naty reports that the Ari, who were dominated in the early period of the Amhara conquest, represented themselves as powerless people: as men 'becoming women'; 'becoming sheep'; 'being castrated'; experiencing 'the shortening of the penis'; or 'lacking a penis' (Alexander Naty 1992: 257). In contrast, the Banna never presented their self-image as female, and labelled the Ari as 'tired people' or 'people accepting defeat'. The Banna said 'the Ari never receive guns for bridewealth' and 'the Ari prefer money instead of guns'. A gender dimension is present if the continuation of resistance equals violence and masculinity, and if the suspension of resistance equals subjugation and femininity. Violence was the self-representation projected by the Banna towards Ethiopia and neighbouring societies, and this image probably corresponded to a stereotype of the Banna also held by others.

In spite of the strength of the masculinity-violence linkage, opportunities for violence have been decreasing over the last few decades. I would like to point to two kinds of symbolization as the reasons why the Banna, nevertheless, possess guns as a form of gun fetishism: (1) guns as symbolic violence; and (2) guns as historical symbol. First, because guns are by nature associated with violence, each man can display his violence without firing. In other words, men acquire status and masculinity just by owning guns, even if its magazine is empty. Secondly, guns are imbued with political meaning, which, at least among the Banna, is based on their historical experience. Although gun models have changed many times over the

---

15. However, I cannot say that guns completely replaced spears as a ritual symbol. While, as I noted in a previous study (Masuda 1997), twin-headed metal spears represent the *Bita*'s ritual power, guns have not become idolized objects of collective memories and worship.

16. See Gilmore (1990) and Cornwall and Lindisfarne (1994). Masculinity should not be portrayed as unchanging. Although I assume that there has been some continuity in attaching value to violence since the age of spears, it is necessary to consider the historical construction of violence and masculinity.

generations, the memories of war, conflict, gun-firing, murder, raids and so on are symbolically accumulated in guns. Moreover, these memories are still being reproduced through narratives. People easily refer to and interpret the memories of violence from the names of old guns during a conversation: the guns among the Banna exist, therefore, as a form of historical index. [17]

17. I am indebted to Dr Taddese Beyene, Professor Bahru Zewde and Dr Abdussamad Ahmad as directors of IES. My research was supported by a grant from the projects 'Comparative Studies on Agricultural and Pastoral Societies in Northeast Africa' (1993–94), 'Comparative Studies on Indigenous Knowledge on the Environment in Ethiopian Societies' (1998) (project leader: Katsuyoshi Fukui of Kyoto University) and 'The Role of Modern Education and Literacy for National Unity in the Periphery of Post-Socialist Ethiopia'. These projects were funded by a Grant-in-Aid for Scientific Research and Grant-in-Aid for Encouragement of Young Scientists from the Japanese Ministry of Education, Science, Sport and Culture.

This chapter is a revised version of my article which was originally published in Japanese in *Minzokugaku-Kenkyu* (Japanese Journal of Ethnology, vol. 65, no.4, 2001).

Chapter 3

Modernization in the Lower Omo Valley and Adjacent Marches of Eastern Equatoria, Sudan: 1991–2000

Serge Tornay

Addressing the question of changing identifications and alliances over the last ten years, that is, since the fall of Mengistu, in the remote south-west of Ethiopia and adjacent marches, or borderlands, of Sudan does not make sense without first describing the background.

Are there 'more chances on the fringe of the state'? It is under this title that, at the Bergen workshop of April 1992, I addressed the question of the 'growing power' of the Nyangatom during a period of about twenty years since the beginning of my study (1970) and the beginning of humanitarian assistance (1972). I suggested that two factors contributed jointly to their growing power: (1) medical and alimentary assistance, which allowed the population to shift from about 5,000 in 1970 to over 13,000 in 1990; and (2) a high degree of political autonomy. The latter was the case, of course, before 1974, but it was also true under Mengistu, who tried to rehabilitate the '*Shanqillas*' along the Ethio-Sudanese border, and who supported the Sudanese insurgency, the SPLA. In the early 1970s, the Nyangatom had been pressurized by the Dassanetch, their southern neighbours (see Map 3.1). The Nyangatom population numbers had declined and they considered that they had failed politically. Twenty years later, they had become the dominant people of the Lower Omo area. This trend was reflected in my 1992 contribution:

> Compared with the often desperate conditions of millions of Africans, I can at least bring a note of hope and optimism from my recent trip to the Lower Omo. During the 1991–92 dry season, the Nyangatom were at peace with all their neighbours, and in December and January they were harvesting bumper crops of sorghum, both from river banks and irrigated fields. The political atmosphere was nothing less than euphoric. An Ethiopian government had at last conceded to them a large degree of autonomy. In October, 1991, twenty-five young Nyangatom had been trained in an EPRDF camp in Awasa and they had been sent back to their country with Kalashnikovs, as a purely tribal militia, committed to maintaining local

order under the guidance of their elders. It was said that no taxes would be
levied and that political and economic autonomy would no longer be
questioned. Being heavily armed – the twenty-five 'official' Kalashnikovs
were just the tip of an iceberg – safe from starvation, and with a rapidly
growing population, they had many reasons to feel confident about the
future. Their optimism, which I shared, since I had seen them for many
years in a worse condition, might have been exaggerated. Sooner or later
hunger would return; they would have to fight again with their neighbours;
relations with Ethiopia or Kenya would bring new problems; droughts
would come again and crops would fail; the army worm would destroy the
sorghum; diseases would kill cattle and small stock or people. The
Nyangatom knew all this, but they were able to feel confident, provided
these misfortunes, which are their common lot, do not affect them all at
once. (Tornay 1993: 151)

The history of the Sahelian area, during and after the colonial period, exhibits a
common feature: conflictual relationships between sedentary agriculturalists and
pastoral nomads. Owing to a higher military capacity, nomads usually dominated
their neighbours, but, with the arrival of modern states, the sedentary, more formally
educated, populations often gained the advantage. This has often been experienced
as the 'rightful revenge' of formerly subject peasants. It seems that in Ethio-Sudanese
border areas such revenge hardly occurred. As an example, take Jon Abbink's (1993a)
article on the Dizi and the Suri of south-west Ethiopia. The Dizi were trapped
between the imperial expansion under Menelik (where they were submitted to both
the *gabbar* system and slavery) and permanent pressure from their southern
neighbours, the Suri, who never accepted submission. Like the Nyangatom, the
Toposa, the Murle and others, the Suri remained, and still are today, outlaws who
take advantage of their marginalized situation (see also Abbink, this volume). Such a
political setting might be at the root of modern forms of rebellion and warfare
organizations, for example, in the case of the Sudanese People's Liberation Army
(SPLA). Taking advantage of living on the margins could be seen as a truism today,
as so many cases, from Somalia, Sudan and many other parts of the continent,
illustrate. It is more frequently acknowledged that the imported state has proved to
many African people and countries to be a dreadful, poisonous, gift. But of course
this cannot be said of Ethiopia, where the state is ancient and indigenous, at least in
the imperial core of the country.

I first met the Nyangatom in imperial times (1970–74). To them, *Mangist*, that
is the state or the government, was an external reality, either hated or ignored. Locally
it was still a time of traditional 'ethnic' conflicts. Under Mengistu (1974–91), the
world of the local populations enlarged both towards the Ethiopian state – Mengistu
fought for a better integration of the formerly despised and marginalized *Shanqillas*
– and towards wider horizons through hazardous 'openings' in southern Sudan. Of
the populations of the southern Omo Valley, the Nyangatom took the best advantage
of the new opportunities, not only by chance, but because they have been, since
1972, the main, if not unique, beneficiaries of an effective programme of assistance

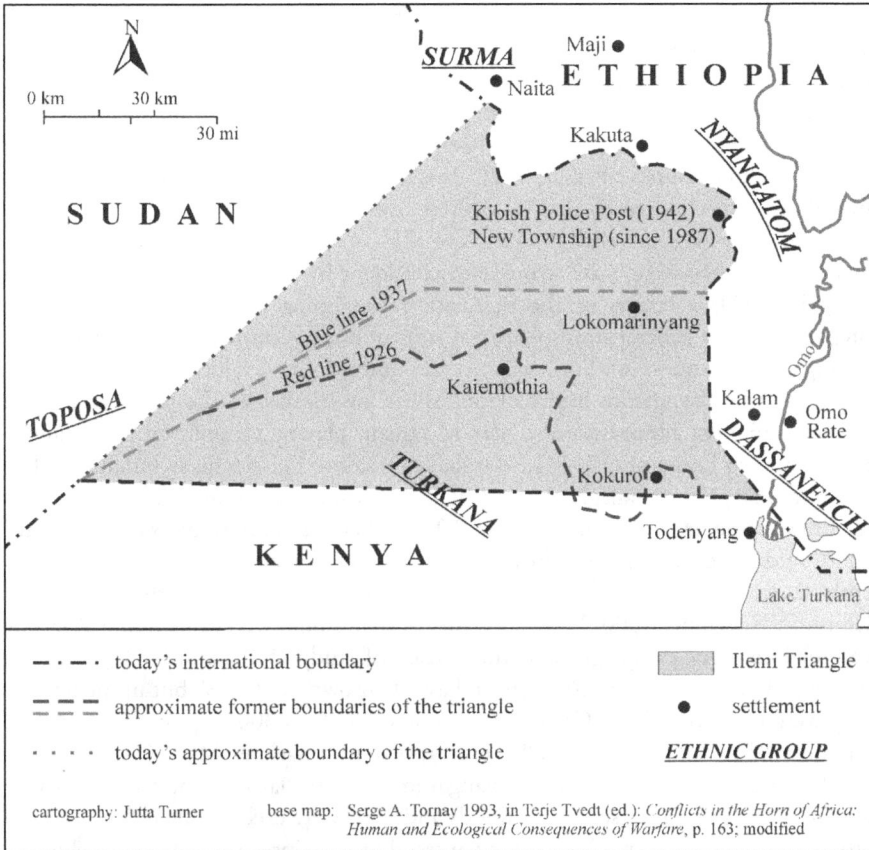

Map 3.1 Map of Ilemi Triangle showing ethnic groups, settlements and administrative stations (source: S. Tornay).

from the Swedish Philadelphia Church Mission (SPCM). Objectively, their chances remain great today, even if they panic at seeing SPCM progressively leaving the scene. Their choice remains open between a soft integration into the Ethiopian zonal administration from Jinka and a freelance, de facto statelessness, by crossing a dry riverbed. At the same time, their great fortune during the 1990s was to have escaped the dangers and hardships of hosting refugee camps: there are no Itang, no Naser and no Kaakuma along the vast territory they share with their Toposa allies, a huge triangle between Narus, Naita and Kibish, an area larger than the Ilemi Triangle. But for how long such marches will remain under their control nobody can say. There is at least one efficient foreign power in the area, the Kenyan army, which continues a progressive annexation of the Ilemi Triangle (Map 3.1).

I agree entirely with Patrick Chabal when he writes: 'Ethnicity has been excessively reified, meaning that it has all too often been studied separately from the more general question of identity. Ethnicity is not an essentialist attribute of the

African, but more simply one of several constituents of identity' (2001: 19). But such a concept is not universal: I heard from a known scholar that, to many Ethiopians, ethnic groups are, like vegetal and animal species, 'natural creations of God'. The cognitive and pragmatic consequences of such a conception of ethnicity are evident.

In the course of the last two or three decades, if the Nyangatom still proved to be the 'Yellow Guns' of their past, they have also been nicknamed by their neighbours 'the children of SPCM', in spite of the fact that, in the course of the last five years, their relations with the Pentecostal NGO deteriorated. In the early 1990s, as I recalled, they were quite strong and confident in themselves. When I revisited them in 1994, a census of the population was going on, elections were being prepared and the general atmosphere was still euphoric because of a joint Toposa and Nyangatom conquest of Suri territory around Mount Naita (see Abbink, this volume). The Nyangatom leader of the SPLA incursion into the Naita area was a former Ethiopian administrator under Mengistu. Having no political future at the fall of the Red Emperor (1991) he decided to become a dissident in Sudan. He thus changed his political allegiance, but without losing one ounce of prestige and legitimacy among the Nyangatom. In 1996, when he came from Naita to recruit child soldiers at Kibish, on Ethiopian territory, he returned on foot with an enthusiastic group of about forty boys from well-famed if not wealthy families. At Kibish in November 2000, I met survivors of that group, who came home after one year of hazardous campaigns and three years of study in Kaakuma refugee camp, northern Kenya. Thus, the Nyangatom have their own 'lost boys', but unfortunately they were not given US or Canadian passports like the 4,300 'happy survivors' who had decided to stay in the camp 'for ever' (press informations, May 2001).

Since 1995, the situation in Nyangatom country has deteriorated for various climatic and political reasons. The floods and rains were excellent in 1994; beautiful crops were growing for the December harvest, but then insects invaded the fields and destroyed the crops. In January-February 1995, returning SPLA officers and soldiers brought bad news from Naita: the Sudanese government had been bombing Narus, the Toposa settlements and the SPLA camp on the border between Sudan and Kenya. Of course, Naita is over 200 km away from Narus, but there was a fear of what the next target of those 'high-flying Antonovs' would be. Women and children were taken from Naita to Kaakuma refugee camp. In February, according to a letter by Joseph Loteng (Plate 3.1, the man to the right), a Nyangatom who worked for the census, a 'sudden reaction between Mursi and Nyangatom along the Omo resulted in the death of six Nyangatom and five Mursi'. David Turton told me later that the event had not been confirmed, but the story is certainly connected with growing tensions between the two 'Nations'. On the Omo, the Muguji 'People' had quarrelled with the Kara 'Nationality',[1] and so they shifted their alliance from Kara to Nyangatom, a point first addressed by Hiroshi Matsuda (1994). The Muguji have now been included in the Nyangatom 1994 census, counted as a total of 417 persons: this is one of the ways a new 'section' can emerge in an existing polity. In

---

1. The 'Southern Omo Zone', with the capital town Jinka, is officially termed SNNP, that is, 'Southern Nations, Nationalities and Peoples'.

Plate 3.1 Nyangatom men in company of an Ethiopian agronomist working for the SPCM NGO (photo: S. Tornay, 1992)

June 1995, a peace meeting was organized by the government, which brought together representatives of 'Bume, Nyangatom, Hamer, Mursi, Kara and others'. Dassanetch were not mentioned, and in August Yvan Houtteman (studying this region in the mid-1990s) informed me that, due to the deadly conflicts of the 1970s, the relations between the Nyangatom and the Dassanetch were still very hostile: 'When I mention the Bume,' he writes, 'my Dassanetch friends point their fingers to their chest, showing where the bullets should pass' (August 1995). There was a vast no man's land north of a line from Omo Rate to Kalam. If there was no real clash between the Nyangatom and the Dassanetch, there were some between the Dassanetch and the Turkana. Four Turkana were killed in revenge for the killing by the Kenyan army of two Dassanetch men who went to Turkana country 'for buying blankets'. 'When I visited Kangaten,' Houtteman writes, 'I got the impression that the Nyangatom are much better armed, in quality and quantity. The Kara also fear the Nyangatom' (21 August 1995). There were good reasons for fearing the Nyangatom: early in 1993, six Nyangatom had been killed when they came, on board on SPCM vehicle, to bring medical assistance to the Kara. The victims were Louryen, the son of the late Lokuti Nyakal, the greatest Nyangatom leader during the 1970s and 1980s, his wife, children and friends. Still today this absurd murder – allegedly for a stolen gun – is not forgiven. Various other incidents could be quoted, but they would not change the general features that still show a very 'ethnic' pattern of relations in the area in the last decade of the last century.

In sum, in terms of conflicts and alliances with the immediate neighbours, nothing very special happened, apart from the conquest of Naita and the eviction of the Suri owners of that country further north towards Maji. The Suri, called Ngikoroma, have always been considered resolute enemies by the Nyangatom. The exact opposite relationship holds between the Nyangatom and their 'grandmother's thigh', the Toposa: no war is reported between them, and that is why their alliance in Naita is so strong against their common enemies. Between the neighbours in the Lower Omo, if Karl von Clausewitz's famous statement applies that war is a 'continuation of politics by other means', it seems equally relevant to say that, to the peoples of the Lower Omo, 'peace is nothing but war conducted by other means'. In this domain, the Nyangatom and their neighbours seem to be very conservative. Periods of peaceful relations are interrupted by 'a sudden reaction', which might be caused by the theft of a gun or of grain, a rape, an injury or a murder, which call for an immediate response. Among some documented examples, I shall mention one that occurred in January–March 2001 between the Hamar and the Nyangatom: a Hamar shot his Nyangatom personal friend in the back while both were apparently in peaceful company in a riverside garden. The victim survived but the Nyangatom carried out four consecutive cattle raids of growing importance. The legal authorities of the zonal government in Jinka do not seem to be able to enforce peace. Today, a Hamar has succeeded a Nyangatom as president of the zone, but this has definitely no link with the recent conflict. It is necessary to underline that non-resident Nyangatom, whatever their positions in the administration, are perceived very negatively in the country. Nyangatom elders consider their sons, once elected to administrative jobs in Jinka, to be lost if not treacherous children. They accuse them of 'driving for themselves' and of playing the game of former Amhara predators. The only modern leaders whom the Nyangatom accept, because they share a common life, are the locally elected heads of *kebele* or localities, initiated under Mengistu and not abolished by the present regime.

To summarize, 'One misfortune never comes alone.' As we have seen, during the 1990s, which initially were years of plenty, the pastoral part of Nyangatom society gained in power: human and animal demography increased considerably; Suri or Surma people had been expelled from their Ethio-Sudanese borderlands, especially from the Naita area, and chased away to the north. The Nyangatom took Naita as their stronghold, joined together with Toposa herders and warriors there and developed an SPLA base. It might be that such a move to the west and to the north – a continuation of the famous drift on which we elaborated already in the 1977 Osaka symposium – had become a necessity: Nyangatom herds had overgrazed the whole of the Kibish area and the grasslands between Kibish and Kangaten, where a joint project (Norwegian Church Aid (NCA)/UNICEF/SPCM) had built rainwater harvesting plants with cisterns of 30,000 litre capacities. Water availability enabled long periods of residence and hence overgrazing. Zebus, goats and sheep multiplied, but donkeys did so excessively. Of course, donkeys are useful as pack animals, expecially on the 100 km between Kibish and Naita where there is no motorable road, but donkeys eat up not only the grass but also its roots. In November 2000, two inedible species (*Tribulus cistoides L.*, Zygophylaceae, and *Heliotropium steudneri* Vatke, Boraginaceae) had almost completely replaced the grass cover between Kangaten and Kibish; only herds of donkeys were still around; cows had left the

Kibish (Nakua) area, once the economic and political centre of Nyangatom, for Naita. What Marshall Sahlins (1961) said about the Nuer here holds true for the Nyangatom: either they disappear or they become 'a system of predatory expansion'. Sahlins attributed that quality to a segmentary lineage organization, but the same applies here, I believe, to the generation system of the Nyangatom, the Turkana, the Toposa and related peoples (Tornay 1995).

As could have been predicted ten years ago, the general atmosphere in Nyangatom is no longer euphoric. When I revisited them after the Addis Ababa conference in November 2000, I found them psychologically depressed and highly pessimistic. The first and universally acknowledged reason for this was that their benefactor for the last thirty years, the Swedish Philadelphia Church Mission, was in the process of 'abandoning' them: the clinic and dispensaries had already been handed over to the Ethiopian government. The consequence of this was that no medicine would be given for free any more, and in practice hardly any medicine would be dispensed in the area. Some assistance was still extended in the name of SPCM, but no one could say for how long this would continue. At the time of my visit, a huge food relief operation was also going on. In response to two years of crop failures and the ensuing starvation, European and Canadian wheat, maize grain and flour were distributed on a large scale. Vehicles of the Ethiopian government transported the food from Arba Minch to Omo Rate, and then back up to Kangaten, the SPCM post on the eastern bank of the Omo. The Ethiopian government, having no vehicle of their own in operation on the west bank of the Lower Omo, not even a military vehicle, simply requisitioned the SPCM to take food across the river. The SPCM had to distribute food locally and transport what was needed to the Kara, Muguji, Mursi and other peoples in the area. They had to do this for free; not a cent was paid for fuel. The SPCM workers from the highlands were upset and somehow agreed that the whole project of the Pentecostal NGO could not continue to be operative. Schooling had also decreased: in contrast to the past, only a vague primary and some grades of secondary schooling seemed to operate. Moreover, no 'educated' Nyangatom had yet agreed to teach in Nyangatom country. According to news in May 2001, all local schooling had been transferred to the regional capital of Jinka, five Nyangatom having finally agreed to teach there. The only Nyangatom who have jobs today outside Jinka are a small crew of SPCM workers: gardeners, mechanics, *zabanya* night watchers and one health assistant.[2]

What about agricultural development? The main objective of SPCM, was to encourage irrigated agriculture and sedentarize the pastoral population along the River Omo. Various attempts to implement solar energy pumps to carry water from the river finally proved to be failures: the wonderful solar panels of 1991 are now used to hang out the washing or to dry crops. Fuel pumps are more reliable, but the cost of the fuel – together with the cost of transporting it almost 1,000 km – makes

---

2. Henceforth paid by the government but not working because of the lack of medicine, this man had become a rich small-stock owner because he gradually invested his wages in buying goats and sheep.

its use prohibitive, unless under foreign sponsorship. The flooding of the River Omo itself is also highly unpredictable: for three years (1995–98) the flooding was so extensive that irrigation around Kangaten was abandoned and the channels neglected. Over the last two years (1999–2000) the river has not flooded over its banks. Since no irrigation had been carried out, starvation began in the country. Food shortages and suffering were finally followed by the food relief operation of November and December 2000. The Nyangatom may have increased in number in recent years – I was told that their demography had reached 20,000 today, but it is difficult to check such a figure – but evidently their dependency, especially on food and medicine, has increased correspondingly.

In thirty years of SPCM development in Nyangatom, the geographical centre of the Lower Omo area, no regional, that is, inter-ethnic, market – with the exception of the one in the new town of Omo Rate in Dassanetch country – emerged between the peoples of the area, the Hamar, Karo, Muguji, Mursi, Nyangatom, Turkana and Dassanetch. Only some forms of dualistic barter survive in some contexts (Nyangatom-Hamar, Nyangatom-Kara, Nyangatom-Muguji, Kara-Dassanetch and episodically Hamar-Dassanetch). In spite of the introduction of currency, mainly through wages paid by the NGO, the failure to generate new, regional, ways of sociability, is patent.

The absence of a regional market cannot be attributed only to possible shortcomings on the part of the NGO. For sure, SPCM agents did not spurn material concerns, but they did show some protectionist or monopolistic attitudes. It is true that they have been, and still are, the only agency with vehicles on the west bank of the Omo. Modern transportation of goods is, then, a de facto monopoly. Why should we exclude the people's accountability for the non-emergence of markets,[3] a social and economic form of activity that is universally cheered in Ethiopia, as it is elsewhere in Africa? Among the factors that put obstacles in the way of the emergence of a regional market I would quote: a very low population density – probably no more than one inhabitant per square kilometre in the whole area; very long distances to walk to meet people of other 'nationalities'; a very hot climate, and a lack of water supplies on the way; lack of information about 'what is going on there'; permanent fear of being 'raped or killed by our enemies'; and finally, of course, the absence of a regional political authority – rare peace meetings do not suffice for creating a constructive political consciousness.

There is one domain in which the Nyangatom have acquired a real degree of independence, and that is religion. The Swedish missionaries were and are Pentecostals (Tornay 1997). They have not themselves been proselytizers of the

3. Instead of developing a regional market, Nyangatom employees waste their Ethiopian money in 'Kibish City', a new town on the Sudanese side (today Kenyan by annexation) of the River Kibish, to the Eastern edge of the Ilemi Triangle (Map 3.1), where Somali and other merchants operate. There, an addicted Nyangatom would pay almost the equivalent of his monthly wage for a bundle of chewing tobacco. The same was extorted thirty years ago by the Kibish imperial police! One can only wonder why the Nyangatom accept such unequal terms of exchange today. Is it through lack of individual entrepreneurship or because of crystallized cultural bias?

Plate 3.2 Pentecostal service (photo: S. Tornay, 1994)

Christian faith, mainly because of pressure from the imperial and then the Marxist governments, and moreover from the Ethiopian Orthodox Church. During the last half-century, Reformed and Catholic churches developed as (almost) purely Ethiopian churches in many areas. The Nyangatom created their own Pentecostal Church under the well-known label *Hiwot Berhan* ('Life Light'). Today they have two modern and handsome church buildings, one in Kangaten and the other in Kibish, the construction of which was financed by Swedish Pentecostals and Canadian Adventists. 'Ethiopian *Hiwot Berhan*', said the Nyangatom, 'tried to eat up the money for building our churches. We refused and all the money has been invested in our buildings. Now of course we have our independent Nyangatom church, but no single cent from outside! We are abandoned!'

The church service that I had the opportunity to witness showed that a new Nyangatom culture is emerging through religion, and that a strong element in it is a new, artistic – in singing – and theologic – in praying –, role of women (see Plate 3.2). Through this 'revolution' a new identity may emerge, either within Nyangatom society or by dissidence. The 'believers' (in Nyangatom *ngi-ka-nup-ak*) do not stand too firmly against the 'traditionalists' or non-believers; in times of crisis the converted say, 'Let us live together and share everything like the early Christians.' Saying this, they wonder why the SPCM missionaries, overt believers and more evidently rich, do not share or give away for free all the goods they have. When discussing SPCM's departure, they say, 'It is not their money which is finished, it is their heart: they have lost their heart.'

For the present, modernization has not yet deeply affected the society. Transformations are superficial as the general conditions of life demonstrate: the customary clothing and other bodily attire, the huts and kraals, the food habits, the whole lifestyle are well preserved. At the same time, it seems that the relations with the neighbours still operate in a rather traditional, 'ethnic' way. An alternation of sudden warfare and fragile peace remains the normal mode of communication between the peoples. Nevertheless, a root for a potential *sentiment de solidarité* does exist: the inter-ethnic institution of personal friendship, which is somehow considered sacred, and does save lives in clashes or other hostile encounters. But a political, regional solidarity has not yet emerged, except possibly in Jinka, between the pluri-ethnic members of a common administration[4] and more specifically between secondary school students who are sharing new modes of livelihood day and night. In this context, new types of relations are also emerging between the sexes.

Let me conclude on the recent conflict between the Hamar and the Nyangatom. My correspondent writes, 'The Nyangatom are not interested in returning the stolen cattle' (Addis Ababa, April 2001). The details clearly show that *Mangist*, the state and/or government, although it exists and has members from the various peoples of the zone, is still a foreign reality. Jinka, 120 km to the north, remains unknown to a majority of the people of the Lower Omo valley. The video record of the public audience given to me at Kibish in November 2000 is particularly significant. To the Nyangatom in the field, there are 'more chances on the fringe of the state': if women today believe that there is no spiritual salvation outside their church, men continue to think that there is no military or food security dispensed by the Ethiopian *Mangist*. They continue to perceive *Mangist* to be a foreign predatory agency, if not a public enemy. Evidently, the expected African Renaissance will not primarily rely on state structures, but will require, especially from formerly marginal or marginalized populations, a new, wider understanding of political action.

---

4. But of course negative feelings also emerge: the Nyangatom elites in Jinka describe the Ari men (the main local population) as 'abominable wizards, drunkards killing each other with knives and spears every Saturday night. On Sunday mornings, the alleyways are filled with their mutilated bodies.' I walked through 'the town' one Sunday morning and could see no evidence of my friends' depreciative statements.

Part II
Institutions of Identification and Networks of Alliance among Rift Valley Agriculturalists

Chapter 4

Burji: Versatile by Tradition

Hermann Amborn

Burji Today

In this chapter I would like to discuss processes that have taken place in turning a community of D'aaši speakers into a people who now consider themselves to be Burji. The evidence of these changes is drawn from seventy years of ethnographic research, some of which is my own. While belonging to the D'aaši community in former times provided a taken-for-granted framework for the identity of the individual, Burji people today have multiple identities. In this context, awareness of being Burji plays an important role. Therefore, my chapter is not about radically changing or giving up one's identity, but its multiplication and transformation and the awareness of where one belongs. I shall try to describe some aspects of this development and their consequences for the people concerned. I shall concentrate not so much on the nature of external political and economic factors, as these are well known.[1] Instead I shall concentrate on the way in which the Burji attempted to come to terms with these factors. These external factors have generally had a more direct and immediate effect on the settled Burji than on mobile pastoralists in the region.

I start with the present-day situation of the Burji. Currently the Burji are a complex social formation, which is internally differentiated and also spatially dispersed within two modern states, Ethiopia and Kenya. Despite the great heterogeneity of these communities, there are close bonds and networks, expressed especially through the offering of mutual assistance to those in need, both within and between settlements. The original homeland of the Burji is situated on the eastern side of the Ethiopian Rift Valley in the southern part of the Amarro mountains. Today about two third of them still live there and some villages have been completely abandoned during the last thirty years. Emigration started slowly at the beginning of the twentieth century with movements of Burji southwards (Amborn 1988). Nowadays Burji communities can be found at various places along a line reaching from Addis Ababa in the north, through southern Ethiopia and northern Kenya, to Nairobi and Mombasa in the south (see Map 4.1). There are large Burji communities in, among other places, Agere Maryam, Mega, Moyale and Marsabit (see Kellner, this volume). These communities have developed in different ways: the areas of settlement are quite different, and the communities have chosen different linguistic and religious affiliations. In some diaspora communities the majority of the Burji no

---

1. Donham and James (1986); Clapham (1988); for the Burji and neighbouring groups see Amborn (1988).

Map 4.1 Location of Burji, Konso (with *Fuld'o* artisan network) and Gamo.

longer speak their own language, despite frequent efforts to revitalize it. In the north they speak Amharic; in parts of southern Ethiopia and northern Kenya they speak Boran Oromo; further south they speak Kiswahili.

In terms of religion, the Burji are present among the followers of all the main formal religions in the region. Muslims are particularly numerous (especially since the 1930s), but in more recent times evangelical Christianity of various denominations has become popular. This has not necessarily been concomitant with an abandonment of the old religion: in some circumstances, the traditional religious ideas are invoked and even combined with the new formal religious ones. Woche Guyo (a Burji from northern Kenya who installed his own archive on Burji history) described how 'Burji Muslims in Marsabit and Moyale are at the forefront of the Tabliq ([Muslim] Crusade) … However [they] are still governed by their traditional religion, especially in ... death rituals ... It is not surprising to see a Burji Haj make secret consultations with the seers' (letter from Woche Guyo, 12 September 2001). Special Burji dignitaries still receive instruction in the traditional religion and belief system, and are considered to be guarantors of the order of the world.

The livelihoods of the Burji also vary greatly from place to place. In their original homelands agriculture is still predominant, but in more urban environments Burji have been successful in securing middle-income occupations. The Burji also play a significant role in trading, especially long-distance and wholesale trading. In some small towns along the road from Addis Ababa to Nairobi, almost 90 per cent of the Burji men living there have been active traders at some time in their life. Some of these people have achieved considerable wealth through trading, and it is striking that the various Burji communities have a higher standard of living than their neighbouring groups. In this respect, parallels are often drawn between the Burji and the Gurage, who have a reputation for being highly successful traders.[2]

It is certainly problematic to describe this social complex, with its wide spatial and socio-economic differences, as a single ethnic group. But, despite their great heterogeneity, the approximately 100,000 Burji, scattered over a large area, still consider themselves to be a unit. This is notwithstanding the fact that other (non-Burji) identities may be equally important for the individual (Amborn 1998: *passim*). I propose here that this shared identity is not so much solely due to external threats, but is related more to the way in which Burji identity is strengthened through the creation and maintenance of new networks. These networks rely on the active reaffirmation of ideas of identity. Thus, the Burji are not among those who feel that they must resort to force and wield a Kalashnikov in order to assert their identity.

Roots of Flexibility

This situation leads one to ask: how is it that the Burji, in comparison with other groups in this geographical area, can cope better with the challenges of the modern world and moreover maintain their identity despite their dispersal and fragmentation? I propose the following working hypothesis: cultural flexibility in former times offered the people a wider range of appropriate reactions to new situations, and therefore a degree of cultural resilience.

Until shortly before the end of the nineteenth century, the Burji did not experience any large-scale external interventions of the kind that have characterized more recent times. Over a period of about 300 years, the D'aaši-speaking people lived as expert hill farmers on the eastern side of the southern Ethiopian Rift Valley. The most striking features of their agriculture were the carefully built terraces on the hillsides, which were combined with irrigation. With their highly developed agriculture, the Burji, along with the Konso, D'iraša and Dullay, were (and are) among the most successful and experienced farming populations in north-eastern Africa (Widgren and Sutton 2004). This intensive agricultural system (called an

---

2. However, the Gurage living between Dilla and their homeland have a wider network in the national trade than the Burji.

'agrarian intensification complex') is by no means rigid. Quite the contrary: the unpredictability and variation of the natural environment led to flexible management of resources, based on long experience. The combination of many different techniques and crops ensured that the agricultural cycle could be adapted to different situations (Amborn 1989). For example, due to the use of thirteen varieties of sorghum, at least sixty different cultivation methods were available. Thus, a broad scope of action had been prepared; appropriate action or agricultural areas could be selected to suit particular conditions. This was essential for survival, especially in recent times, when the ability to react flexibly to external events has proved important.

In addition to scattered settlements and a few small villages, they lived in urban settlements, the biggest of which in Ethiopia, Boohee Burji, was an important supra-regional trade centre, with about 8,000 inhabitants in 1955. This settlement may have grown up partly to serve the trade that passed through this area in the nineteenth century (and even earlier). Boohee Burji was, at least at certain periods, the end point for caravans from Somalia.[3] This was partly due to its favourable geographical position. It lies on the south-eastern boundary between the two types of farming: settled cultivation and nomadic livestock-keeping. Despite its mercantile significance and its adaptation to serving the passing traders, it continued to be mainly an agricultural settlement (Amborn 1995).

There are other cultural areas marked by great flexibility. I shall just mention the main areas here. The most important is social structure: the history of the Burji shows clearly that, as with most groups in south-western Ethiopia, they are not and never have been a tribe in the classical sense. Neither spatial nor social boundaries – over long periods of time – were ever rigid. Although they clearly distinguished themselves from some neighbouring groups such as the Guji, their history has in general been shaped by processes of social fission and fusion. Alliances and aversions were constantly changing. This is exemplified by the relationship between the northern and southern parts of their homeland. In the past there have been serious conflicts (Amborn and Kellner 1999: 11). The last bloody conflicts were due to a rivalry that developed between the temporarily elected Burji political leaders (*woma* in the south and *dayna* in the north). These local indigenous leaders could not be installed without the ritual assistance of the other and they were therefore part of a dual complementary system. Their rivalry had developed as they competed for the support of the northern Ethiopian government. This conflict smouldered for about a hundred years and only ended in the last decade following the death of the last of the former *balabbats* (and agent of that northern government).

The above example indicates serious differences within the Burji area, but for non-Burji people in the region the northern and southern groups are seen as one unit. The term Burji became accepted as a common name for the whole group. The

---

3. Sasse and Straube (1977), with excerpts from Straube's manuscript of his first fieldwork in 1955 and from his 1973/74 field notes. These notes are also the basis for Amborn and Kellner (1999). For the use of the terms Burji and D'aaš (D'aaši), see Amborn (1995: 75).

relative and blurred nature of local identity can also be seen in the case of the Koyra (or Koorete). The Koyra are the Omotic-speaking northern neighbours of the East-Cushitic-speaking Burji. In some situations, the Koyra also call themselves Amarro. In some cases, they lend their name to Burji people living outside their core area. For example, the Burji potters in Konso call themselves Koyra. Identity is stressed according to what appears to be politically expedient. When *woreda* boundaries were drawn up recently, the inhabitants of the southern Koyra districts (Naddalley and Haralle) voted to belong to the Burji Special Woreda, and hence stressed their common attributes with the Burji. But this move did not go uncontested by some Koyra and the disagreement led to conflict.[4] The fluidity of identity and the way in which it can be manipulated are further illustrated in another case: when the Burji living in Marsabit experienced difficulties following Kenya's independence, they turned to Haile Selassie for help. In their letter they referred to themselves as Burji-Amarro and claimed a relationship between Amarro and Amhara, based on popular etymology. Having thus mutated to Amhara, they begged for his support, and declared themselves to be the Emperor's lost children. This strategy proved effective.[5]

Within the northern and the southern parts of the Burji homeland, there has been fusion and fission in the clan and lineage system. While the other ethnic groups in the Burji-Konso-Dullay cluster each have nine patriclans, the Burji, despite structural agreements with their neighbours, have a great variety of clans and lineages, although separation into clan and lineage is in some cases not discernible. Nevertheless, ideally, the number nine predominates. In my opinion this is an indication that processes of fission and fusion among the Burji played a greater role than among their neighbours. Looking back over longer historical periods, it can be seen that the process of group formation, including the discussion of social concerns, has never come to a standstill.

Alliances also exist that stretch beyond the areas settled by the Burji and Koyra. First, clan correspondences should be mentioned, in which a clan among one ethnic group is said to have a corresponding clan in another group, and this is used as the basis for an alliance, a frequent occurrence in Cushitic areas (Schlee 1994a [1989]). In drawing on these alliances, the lineage or clan elders (*ganni*) of the Burji should be mentioned.[6] These have religious connections, based on myth, to corresponding

---

4. Oral communication from Alexander Kellner. Before the boundaries set up by the imperial administration, the districts of Naddalley and Haralle, with a Koyra-speaking population, belonged to Burji territorially and politically, with their own *dayna* (see also Straube 1963: 76). Some Ammarro from the Amarro Special Woreda encouraged the people in these districts to oppose affiliation with the Burji Special Woreda. In his letter from 12 September 2001. Woche Guyo notes: 'There is cease fire now but the situation remains hostile.'

5. Document in the private archive of Woche Guyo, Marsabit. In his letter of 12 September 2001, he adds: 'Haile Sillassie ... wanted all tribes to declare loyalty to him ... Also the Burji had cast their lot with the Italians ... What the better way than to declare the Burji are Manze (Amhara royal family)! It worked.'

6. For the importance and function of the *ganni*, see Amborn and Kellner (1999: 13 and s.v. *ganni*).

dignitaries of neighbouring groups, for example to the *poqalla pamalle* of the Konso and the *d'aama* of the D'iraša (Gidole) (Kellner, this volume).[7] Further social relationships, going far beyond the original group, could also be established on the basis of the generation group system. This was the case with the Konso, and especially included the Boran *gada* (generation-grade) system.

Forms of alliance were also important between individuals. At least until my visit in 1980, partnerships were frequent between a cultivator and an ally in a neighbouring pastoral society.[8] Each undertook to provide the other with assistance when needed. Such partnerships could be based on clan or *gada* relationships (as described above), but alliances also existed independently of these systems. The symbiosis between cultivators and herders was a stabilizing factor for both forms of subsistence, especially in times of crisis, since a period of drought does not necessarily affect cultivators and herders at the same time or with the same degree of intensity. Individual relationships within this network could be suspended for many years, and reactivated in time of need.

Craftsmen were also particularly important in building relationships between groups. Craftsmen were full-time specialists, who earned their livelihood by selling their products.[9] They were more mobile than cultivators, and often went to live for several years among pastoralists (Amborn 1990: e.g. 67f.). The demand for metal objects, including ceremonial objects of the Boran, for example, was met by smiths from Burji, Konso or D'iraša.[10] Up until the middle of the twentieth century, there was also a close relationship between the potters in Konso and Burji. Today this is insignificant, but Konso potters are still considered to be from Koyra, which in this context is understood as meaning to be of Burji descent. Another connection was formed by the weavers, via the islands in Lake Abaya into the Gamo highlands (see Tadesse, this volume). But, perhaps most importantly, trading among the agricultural societies mentioned above, including long-distance trade, fell within the field of activity of professional craftsmen. It was these craftsmen who travelled between areas, brought new goods and were a vital factor in friendly relations with people outside their home area (Amborn 1990: chap. III).

Viewing the spiritual world of the Burji shows us that it is not regulated canonically. There is no 'traditional theological system', but a great variety of syncretisms of different origins, in which each element is as true as the rest. Creation myths in the Burji-Konso cluster reveal a world that is not yet completed. It is the task of humans to continue shaping it and to contribute to the dynamic development

---

7. The connection with the *d'aama* stretches further into the Dullay area.
8. In 1980, my information was from Dullay, Konso and Burji. After 1980 I did not raise this question again.
9. Social division of work into cultivators and different crafts took place during the seventeenth century (Amborn 1990). This dating results from a correlation of oral traditions, genealogies and chronological cycles of the *gada* systems of different ethnic groups within the Burji-Konso cluster (Amborn 2006: 57).
10. From the mid-1920s onwards production shifted away from the Burji towards the Konso and D'iraša.

of culture. Not only are contradictions allowed, but they do not cause any intellectual problems.[11] To give just one example: in order to distinguish themselves from the Amhara, large numbers of Burji became Muslims during the Italian colonial period. At the same time they revitalized the *gada* system, together with its spiritual dimensions. This open system, with unfinished creation and many syncretisms, could be described as a 'discursive religion'. The structural principle on which this conception is based has a direct counterpart in the openness of the permanent discursive social process.

Thus, diversity and negotiability can be seen to be a characteristic of the main cultural areas of Burji. In agricultural production, in the clan system and also with regard to the generation-grade system, there were openness and flexibility.[12] There are counterparts of these phenomena in other ethnic groups, particularly in the Burji-Konso cluster, including the agricultural diversity among the Konso or Dullay. But the important point is that in Burji flexibility and diversity occur simultaneously in all the main areas of life, and they mutually reference and reinforce each other.

Changes, Challenges and Chances

We can now turn to the question: what caused the changes that took place in Burji and how far did their internal history and Burji traditions shape the directions of this change? The first serious change was brought about by the conquest of the northerners: this led to the subjection of the Burji (and many others) to a harsh *gabbar* system and also changed many trade routes. This must not be thought of as an abrupt event, but as a change that occurred in phases over the course of a frontier era lasting for several decades (Amborn 1988). During this time the Burji town lost not only its importance as a central marketplace, but also important trade partners, due to the imperial monopoly on coffee. Since the 1930s, therefore, the Burji have grown coffee only for their own use. Instead, they intensified weaving, which became almost a kind of national craft until its decline at the end of the 1960s.

It was the combination of these two political and mercantile factors that formed the conditions for further development. An additional factor was the active response of the Burji to outside influences, an outcome of their culture-specific flexibility – including the possibility of cultivating variable alliances. Instead of passive adaptation, they successfully made selective use of the opportunities available to them in these new times. Their political and economic situation in the diaspora today is certainly better than that of some other ethnic groups.

The Burji were able to employ their old trading connections, which now, under changed circumstances, they took pains to preserve. They did not organize trading caravans even in former times; Burji craftsmen-traders travelled either alone or in small groups. This was possible because, among the Boran, craftsmen are seen as

---

11. Straube (o.J., c.1955, chap. III); Amborn (1990, chap. II.3, IV.2.2, V.5; 1995: 81, 90; 1998: 356); Amborn (forthcoming).

12. The Burji were not satisfied with just one generation-grade system, they practised two different parallel systems (see note 15).

sacrosanct persons and must not be killed. Previously, the personal alliances of the Burji along the long-distance trading routes may not have stretched further than about 200 km to the south-east, but now these connections, which had been kept up over decades, could be expanded. Burji smiths had been working in some of these more distant places for generations. They served as a port of call for fellow craftsmen-traders. The craftsmen-traders were followed by other Burji cultivators, and this led to the founding of outposts. The cultivators were able to use the clientele of the craftsmen to develop further alliances (see Amborn's chapter on Konso craftsmen, this volume). In the course of time, the cultivators succeeded in breaking down the exclusive position of the craftsmen in trade. Especially in bigger towns, the professional craftsmen-traders lost their significance as a social group.

It is possible to reconstruct this development in the case of Marsabit in northern Kenya. First, in order to escape the *gabbar* system, some individuals or small groups settled in the Boran lands south of the River Sagan (for example, Teltelle, Hidda-Lola), where they worked mostly as client herdsmen to the Boran. Later, some Burji moved further to Moyale, where they established a diaspora settlement. They started to cultivate there so that one commentator remarked that 'the country round Moyale is becoming a vast shamba owing to the activities of these people'.[13] After contacts had been cemented in Moyale, temporary contacts with the Marsabit Boran were formed, followed by contacts with the British and Somali.[14] As a consequence, some Burji families went to settle in Marsabit, where they subsequently made efforts to expand their economic and social network, building on clan and *gada* relationships with the Boran as they had elsewhere.[15] Building the relations with the Boran was helped by the fact that the Burji claim to have a common origin with the Boran, in Liban in southern Ethiopia.[16] In order to reinforce the relationship with a Boran person, *gada*-set membership was also sometimes manipulated, with the use of ceremonial purification rites (the most recent example of a change known to me occurred in 1992). The Burji in Marsabit assured me that, in order to demonstrate their togetherness with the Boran, they also adopted their language. Today, at the

---

13. Letter to the Colonial Secretary, Kenya Colony, from Officer in Charge NFD Moyale, 6 July 1922. Copy made by Woche Guyo.

14. In the mentioned letter from 1922 the officer in charge asked the Colonial Secretary to give land to the Burji in Marsabit. See also Kenya National Archives: PC/NFD/1/1/2 Annual Reports 1920–22.

15. There are clan correspondences between the Burji and Boran. The Burji generation grading system was based on formal links between two generation-set cycles: *gada* and *hagi*. The cycle called *gada* had more in common with the Boran system than the *hagi* cycle. The *gada* cycle was abandoned but the sets are still remembered (Amborn 2006: 59). Schlee's article (1998b) gives good examples for the functioning of such relations, though the relations of the Boran and Burji *gada* systems were not as close in the nineteenth century as those between Gabra, Boran and Garre.

16. For the Liban tradition, see Amborn (1994: 781); Kellner (2007: chap. DI). A short reference to the Liban tradition of the Boran is also made in Gufu Oba (1996: 118). Woche Guyo writes in his letter of 12 September 2001: 'When relations are strained between Burji and Boran the Liban tradition is invoked. It has smoothened matters in most cases.'

most, only 20 per cent of the Burji in Marsabit can still speak their own vernacular language. A contributory factor to this change may have been that in Marsabit the Burji and Koyra regard themselves as one group and use Boran as the common language between them.

The Boran in Marsabit were urged to cultivate the soil by the British from 1946 onwards. When they tried to practise agriculture in addition to livestock-keeping, the Burji helped them with their knowledge and taught them the methods they had used on their outlying farmlands. From the 1980s onwards, however, the relations between the Burji and the Boran deteriorated (Kellner, this volume). As a result the Burji are currently seeking closer contacts with Gabra and Rendille. A new element in Marsabit is the fact that people are prepared to use violence to assert their position in such conflicts.

Because of the circumstances I have mentioned, diaspora communities were and still are formed mostly far apart from each other, and communication between them was at times difficult or even impossible. Several small and thus controllable local networks were formed by alliances with non-Burji clientele. Thus, the Burji made a virtue out of a necessity. Direct access to a network is available only to Burji living in the respective diaspora community.

This structure is best shown by the diagram seen in Figure 4.1. There are Burji communities (Bu(a), Bu(b) and so on, to (n)) along the north-south chain, with Boohee being still acknowledged as the 'homeland'. Each community has its own network (for example, Bu(a)-N(a)), made up of Burji and their clientele. The network can be formal or informal. At the centre of the network there is a core that usually consists of traders (or craftsmen-traders). In the larger diaspora communities the organization of the network corresponds to that of the 'homeland'. Where Burji networks operate, the Burji try to prevent non-Burji people from other towns from having access to members of their network, thus rendering their inter-town contacts exclusive. Sometimes trading plays no part. In Arba Minch, Ethiopia, the main purpose of the network is to ensure that certain middle-income jobs are tied to a particular group (in this case the Burji themselves). But, when drawing on Burji-Burji ties, they have to take care not to show arrogance, as this might cause tension with other local people. Although the Burji, as a minority, avoid getting involved in

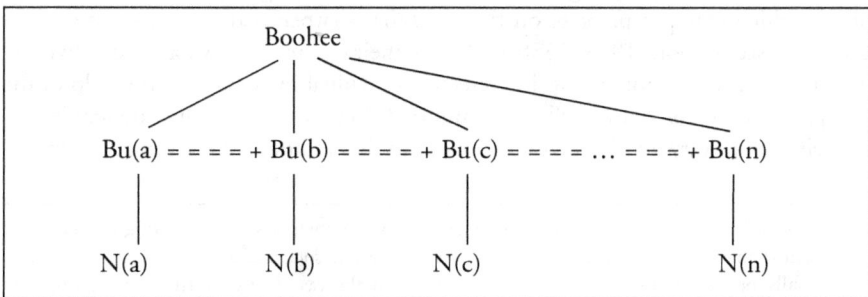

Figure 4.1 Burji networks

disputes, they are not welcome everywhere. This sometimes leads people to deny their Burji origin. But doing this means risking that other Burji will no longer be prepared to give help when required.

For individual Burji it is of great importance that anyone wishing to live for a limited period in another Burji diaspora community can link into the local network there, usually via a member of the same lineage. It is also recommendable not to omit any link in the north-south chain, at least as an intermediate contact. In other words, individuals or small groups can profit from local contacts through Burji middlemen, while the various local networks (a, b, c, ... n) usually have no connection with each other. An example of the integration of young Burji newcomers into the local network is the youth club in Marsabit: members will help in finding a job or they are prepared to collect money to pay school fees.

In some cases situational 'we-groups' are formed.[17] Especially in the communities far distant from Boohee Burji, these we-groups have distinct identities and include people other than the Burji. These we-groups provide individual and collective identity in times of economic and social insecurity. An outside observer would find it difficult to see what a we-group of Christian Burji in Addis Ababa and a we-group of Muslim Kiswahili-speaking Burji in Mombasa have in common.

When a Burji family decides to take the opportunity of moving to another community, it must also be prepared to adopt the identities of any new we-group, and there appears to be little problem with this, which is certainly connected to the fact that fusion and fission have a long tradition in their culture. One example of the way people adapt to the various diaspora communities can be seen in one of my friends who employs various different names, depending on the situation. In a Burji environment he uses his Burji name, Hirbo Borde, together with the initials A. and T. In Meru (central Kenya), and frequently in contexts involving the administration, he calls himself Timothy A.H. Borde. In Somali areas he uses the name Ahmad T.B. Hirbo. As he says himself, with a twinkle in his eye, each time he changes his name he also adopts the expected role.[18]

Now we can turn towards the question of what holds Burji people together in different diaspora communities, despite their multiple identities and great heterogeneity. There is no doubt that a new identity has developed over the past fifty years. Life in relatively small groups, surrounded by bigger groups with different ways of life, which may put pressure on the Burji, has sharpened their awareness of their Burjiness (see Amborn 1998: 355ff.). Despite their openness, they have clear dividing lines in order to promote group interests, or individual interests with the help of the group. Especially for Burji intellectuals and traditionalists, these bonds are manifested in their own history and in their attachment to the traditional homeland area and its

17. These are informal ad hoc groups that give their members a sense of belonging. They are formed mainly in urban environments, where the old social relations of earlier days have usually become weak (Evers and Schiel 1988). In the case of the Burji the we-groups are frequently, but not necessarily, modelled on clientship.
18. Names were changed by the author.

sacred places, even if, for many of them, Boohee Burji has only a symbolic significance.[19] Their rich oral histories are also important (see Amborn 1994; Kellner, this volume), particularly those referring to a common homeland. Ideas about sharing 'the same blood' are also thought today, especially among intellectuals, to be the binding element in this patrilinear society, despite many intermarriages with other ethnic groups. Burji clubs in the diaspora are an important element in the building of identity, and especially all-Burji meetings, to which the clubs send representatives.[20] The merits and demerits of older traditions or their transformations are often discussed there. Although attempts at revitalization rarely have any visible success, the important thing is that cultural questions are discussed. In Konso, where, unlike in Burji, the most significant cultural changes have only taken place in the last fifteen years, intellectuals lament people's lack of awareness of their own culture and admire the Burji, who have made cultural identity a subject of discussion.

The Threefold Capital

In interpreting the situation of the Burji it seems useful to refer to Bourdieu's concept of the threefold division of capital: cultural, social and economic (Bourdieu 1983: *passim*). One sort of capital can be regarded as a resource for acquiring another. The ranges (Bourdieu's 'fields') of those three kinds of capital, however, go far beyond Burji society, and extend even beyond the reach of Burji-centred networks into the wider national societies.

As far as cultural resources are concerned, these depend primarily on the place or places of residence of a person. Burji who do not live in their 'homeland' are always, as they say themselves, 'in the diaspora' if they are English speakers or '*ifatta burjee*' (from *ifa* 'outer zone') if they are speaking Burji. They thus form a minority among people who have different traditions and values. The cultural capital acquired by the Burji includes part of the tradition incorporated by their parents and their social environment.[21] This tradition must not be thought of as uniform; it corresponds to what each Burji community sees as 'its culture'. This area, which is felt to be an essential part of the overall cultural environment, therefore serves

---

19. [The] original Burji territory ... , which today has fallen into insignificance ... but the ceremonial places, the graves of the ancestors, the trees under which the first settlers rested, and all the many parts of the landscape linked with the history of the people and their world view, have lost nothing of their significance ... [today the] cultural landscape around Burji town *as a whole* is given a sacred character. The sacred landscape is not just an idea. Its actuality in practical life is demonstrated, for example, by the fact that Burji who have lived somewhere else for many years wish to spend their old age in the original territory and to be buried close to their ancestors. It is without doubt the ancestors who provide the vital link with the original territory; ancestors ... and locality become one. Time and space are consolidated ... The physical environment suggests permanence, offers a bridge to tradition, and can thus become a symbol of cultural revitalisation and identity (Amborn 1997a: 386).
20. Most of these clubs have a formal structure with a chairman, a treasurer, etc. and organized meetings.
21. On incorporation and the incorporation process in Bourdieu, see Bourdieu (1983: 189).

effectively as a polar force within the cultural capital. The localization and intensity of this cultural capital are neither static nor invariable. An example of this is the Liban tradition, which I have referred to. It is told and assessed in different ways as far as time and place are concerned. More important with regard to the use of resources is that a sense of flexibility and of dynamic processes and discourses is incorporated in the transmission of Burji culture. This knowledge is a cultural resource that can be converted into social capital. The social capital that is used in intercourse with other Burji can be modified and updated beyond the particular group, thanks to the flexibility it contains.

Marriage relationships have an important function in the formation of local networks. In the first half of the last century, there were two established possibilities for young men seeking marriage partners who (as was usually the case) wanted to settle in the neighbourhood of pastoral nomads: the informal, friendly and economic partnership with a herdsman on the part of one of their relatives, which frequently already existed, or institutionalized clan correspondences. If it were not that knowledge of sophisticated agricultural methods or of the rules applicable to trading and crafts was expected to be part of the cultural capital of a Burji, these young men would have stood little chance. After all, agriculturalists are thought by pastoralists to be inferior. But many a father must have thought it could do no harm to marry his younger daughter to a 'soil digger'. A rather loose, friendly partnership would thus be transformed into an obligation on the part of the son-in-law. In times of need he must supply grain or provide help through his trading connections. Here, the use of cultural, social and economic capital flow together.

Today, after several generations of intermarriage, the Burji are no longer alone in the position of having to beg for marriage partners. Marriage arrangements are possible between all local groups. In recent years in Marsabit the alliances strengthened by marriages between Burji, Rendille and Gabra have led even to the founding of a political alliance with the acronym REGABU (= Rendille + Gabra + Burji), which is active in local politics (see: *The Economic Review* 7 April 1997). Burji communities can also use marriage to determine inclusion and exclusion. In-laws, especially in the second or third generation, can be included in the inner network because of their (apparent) Burjiness, or, if they are considered to be useless for the community, excluded, since the relationship only exists via the female line. Here they are following a logic of practice (Bourdieu), without denying the logic of the system.

These remarks require further explanation in order to avoid giving the impression that the Burji are essentially a homogeneous group. Again we must refer to the different kinds of fields of action. Burji in the diaspora are not born into the fields of action there. Rather they have to face pre-existing dispositions. The Burji reaction to these has been doubly oriented: a supra-regional and a local orientation. It is the supra-regional orientation that, although spatially dispersive, comprises the inner Burji field of action, i.e. handed-down cultural, social and economic resources. In relation to the local Burji (who are few as compared with the rest of the population), this field of action is real. But outside the place of residence (in the diaspora) it is rather a symbolic field, whose resources can be activated ad hoc when needed – with a certain amount of personal effort.

The daily environment is the diaspora. Burji living there are the nearest partners who can be related to, but they alone cannot supply the vital resources. Yet other Burji communities are too far away to satisfy daily needs. Thus the second orientation is towards the place of residence, and there it is towards non-Burji groups. A local network is developed, as already mentioned, via extended marriage relationships, via trade, crafts and modern professional connections, in which activities are directed especially towards the creation of economic and social capital. It is no longer enough to rely on handed-down and incorporated cultural capital alone. The medium-sized towns where the Burji prefer to settle today offer good opportunities for acquiring wide cultural knowledge. Due to their limited size, conditions here are also favourable for the creation of social networks, at the same time allowing them to become more open. Here the borders are open: society is in flux.

As far as the identity of the Burji is concerned, we can assume that persons who think of themselves as Burji see their Burji identity as founded in the first place on common descent, and find confirmation of this identity in intercourse with other Burji. The extended and new fields, on the other hand, require individuals and Burji communities to work on building their identity. This does indeed take place within the group, for instance when topics such as the changed social environment are discussed in Burji clubs. The process of identity forming is significantly influenced – to follow Keupp (1999: 198ff.) – by the available resources. It depends on which energies are spent on the selection of potential resources offered by the various types of capital. Only by testing their usability and developing strategies can a process of orientation and giving meaning be set in motion which contributes to building identity.

In conclusion we can say: the Burji people have been able to react in a versatile way to new and harsh conditions thanks to their culture, developed in earlier times on the slopes of the Amarro mountains. Strict codification had no place in the polycephalous social order, where institutions and ideas were always in a state of flux because of their exposure to permanent discursive negotiation. This put many Burji in a position to cope with diverse identities or to change identities, while at the same time building a new identity as Burji. Under the given historical conditions, their culture has demonstrated flexibility and the ability to adapt to new and wider sets of socio-economic conditions and identities by using the manifold potential that is inherent in it.

Chapter 5

The Significance of the Oral Traditions of the Burji for Perceiving and Shaping their Inter-ethnic Relations

Alexander Kellner

Introductory Remarks

In this chapter I shall try to show that the oral traditions of the Burji play a significant role in how Burji perceive and shape their inter-ethnic relations. Myths of origin, migrational traditions or oral historical accounts are not just petty stories that have no relevance for the present. On the contrary, they are important elements of the discursive framework in which people discuss and reflect upon actual problems, their own identity and their relations to other groups. Oral traditions provide, as I shall show, images and blueprints out of which people develop discursive lines in order to evaluate their present situation and to anticipate future possibilities. The outcomes of these reflexive thinking processes then direct people's expectations and actions. Therefore, with the help of oral traditions one has the opportunity to get an insight into people's views on certain problems, such as inter-ethnic relations, and to identify ideological motives that either make people undertake certain actions in regard to inter-ethnic relations or form the necessary ideological base for these actions.[1]

The Significance of Oral Traditions for Burji Identity

Among others, Rüdiger Schott (1968: 170) has pointed out that groups base the consciousness of their unity and specialness on events in the past; in other words, they need the past for shaping their identity as a group. For answering such questions as 'Who are we and where shall we go?' the recollected past is used as a point of reference.

This is especially true for the Burji, who are no longer homogeneous or traditional but are part of a complex and differentiated modern society. Since the

---

1. From October 1999 until October 2000 I conducted fieldwork on Burji oral literature in Marsabit, Kenya (three months), and in Burji Special Woreda, Ethiopia (nine months). This work is part of my doctoral studies on oral literature and Burji culture supervised by Prof. Dr Hermann Amborn, University of Munich. (Date of completion of manuscript: 2002. My PhD thesis has been published: *Mit den Mythen denken. Die Mythen der Burji als Ausdrucksform ihres Habitus*. (With an English Summary). Hamburg 2007: LIT.)

conquest of southern Ethiopia by Emperor Menilek II at the end of the nineteenth century, the majority of the Burji left their homelands and built up communities in various places in Ethiopia and Kenya.[2] The spatial dispersion of the Burji was accompanied by an internal social differentiation into groups with different linguistic and religious affiliations, economic situations and lifestyles. These circumstances made the Burji reflect upon their cultural identity intensively, and they think a great deal about which sociocultural factors constitute 'Burjiness' (Amborn, this volume). Among those elements that are viewed as essential to common Burji culture, a treasure of myths and oral traditions, particularly historical experiences, is of great importance (Amborn 1994: 779).

The Significance of the Liban Story for the Inter-ethnic Relations of the Burji

In this context, the Liban migrational tradition must be mentioned. It is generally in the minds of the Burji and can be denoted as 'the' tribal myth of the Burji. Many Cushitic population groups, such as the Oromo, the Burji and some clans of the D'iraša and Konso, consider the Liban region (in the area around present-day Negele) to be their place of origin.[3] According to the Liban tradition of the Burji, in the beginning the Burji lived together with the Konso and the Boran in Liban. In some versions of this myth, the Burji, the Konso and the Boran are referred to as *mashaana*. *Mashaana* means 'children who have the same father but different mothers'. In other versions, the three groups are referred to as *fira*. *Fira* either means 'relative' in the broadest sense or 'close friend with whom you have a family-like relationship'. *Fira* implies a relationship of mutual assistance.

The story has it that in Liban the Burji, the Boran and the Konso alternately provided a sacrificial sheep or goat each year, which they would jointly sacrifice in order to prevent all evil influences and to assure people's well-being. Some of the Burji with whom I discussed the story said the yearly sacrifice gave expression to the social bond that united the three communities. However, the story also relates that one day, when it was the turn of the Burji to provide the sacrificial animal, the Konso came at night and stole the animal the Burji had reserved for the sacrifice. The Konso ate the sacrificial animal and threw its bones and intestines in front of a Burji man's house. The following day the animal was missing. Since people found its remains in front of a Burji man's house, the Burji were blamed for having defiled the sacrificial

2. The original homeland of the Burji (whose number is estimated to be 90,000) is situated in Burji Special Woreda, on the eastern side of the southern Ethiopian Rift Valley. In the original area, some Burji still practise a highly developed form of agriculture, with the typical elements of agrarian intensification that is characteristic of the whole of the Burji-Konso cluster. For Burji culture and history see Sasse and Straube (1977), Amborn (1988, 1994, 1995), Amborn and Kellner (1999) and Kellner (2001, 2007).
3. For the Burji see Mude (1969: 28) and Amborn (1994: 781). For the Konso see Jensen (1936: 385, 494) and Watson (1998: 291, text 2 (Bamalle)). For the Boran-Oromo see Haberland (1963: 24). Even though various Oromo sections mention different places of origin, Liban is nevertheless regarded as the cradle of Boran/Oromo culture and as 'the' ritual country of the Oromo above all others (see Knutsson 1967: 165; Bartels 1994; Gufu 1996: 118).

animal. The Burji were called by the Boran for several meetings in order to confess but being angry about the fact that they had been wrongly blamed they did not appear. (There is also another explanation why the Burji refused to appear before the Boran when summoned.[4]) Finally, the Boran drove the Burji out of Liban, who then migrated to their homelands, which are situated on the southern slopes of the Amarro mountains. According to the Liban tradition, on their exodus the Burji developed their own way of life and sociocultural institutions. After having reached their homelands, the process of ethnogenesis was completed.

Thus, the myth is exclusive and explains how the Burji separated from the Konso and the Boran and created an ethnic group of their own opposed to others; but it is at the same time inclusive (Amborn 1994: 782). For example, the myth also emphasizes the following: first, the Burji are said to have originally lived together with the Konso and the Boran in Liban; secondly, the myth also speaks of a family or family-like relationship between the three communities; and, thirdly, certain aspects of the Burji and Boran social order are described as being equivalents of each other. In one version, the narrator states that both societies were organized in moieties that corresponded to each other (Jiree Maalloo and D'aashiccaa in Burji, and Sabbo and Goona in Boran). Burji clans are also said to correspond to Boran clans. The Burji clan 'Goodaa', for instance, would be called in Boran language 'Karrayyu'. Such inter-ethnic clan relations are a frequent occurence in Cushitic areas (Schlee 1994a [1989]). These aspects of the myth mean that the Burji see themselves as part of a wider social context that includes the Boran and the Konso, or, as Amborn (1994: 782) has put it, as 'part of a whole'.

If the history is considered further, it can be seen that the process of group formation has never come to a halt. Fusion and fission in the clan and lineage system have occurred, and social boundaries with neighbouring groups have always been in a state of flux. Alliances with neighbouring groups can be sought on the basis of clan correspondences or the generation group system. By referring to the 'whole' the Liban story allows 'the existence of other identities whilst letting a person still remain, first and foremost, a Burji' (Amborn 1994: 783). This point, inclusion of the 'other', must be stressed because in anthropological discussion of *ethnos* and ethnicity it is the exclusion of the 'other' that has been focused on, whereas inclusion of one's own culture has been neglected (Amborn 1994: 782).

The Burji use the Liban story as a discursive medium through which they both reflect upon and shape their identity and inter-ethnic relations. For example, I was a guest at a political meeting in Burji in 2000 where people discussed the revitalization of certain aspects of their traditional political system. During this meeting, bloody conflicts that had taken place in the past between the two Burji branches, the northern and the southern Burji, were recalled by several speakers. Through these

---

4. 'The Burji view is that, though the Burji and Boran are siblings, the Burji have the right of primogeniture. Indeed, the reason why the Burji refused to appear before the Boran when summoned was that they believed the Boran to be the junior. The Boran, on the other hand, say it is vice versa. On this matter, the two have differed to date' (Woche Guyo, Marsabit, correspondence, February 2002).

recollections, some questioned the unity of the Burji. Then, one elder stood up and told the audience that the Burji had come from Liban and had separated from the Boran and the Konso. He went on by enumerating the places the Burji passed on their migrational route, and closed with admonishing people not to question or to endanger the unity of the Burji. Hence, the Liban tradition is used to conjure up and foster internal unity. Also, I was told by Burji in Marsabit that, when tensions or conflicts between the Burji and the Boran have arisen or broken out, people would refer during peace negotiations to their togetherness in Liban in order to support peace agreements. At the beginning of the negotiations one party would say, 'We are all from Liban.' The other party would nod and agree. In this context, the myth also serves as a medium for reflecting upon how the Burji are to behave in case of conflict with another ethnic group. One Burji remarked that the Burji should never again defiantly refuse to take part in peace negotiations, as they had done in Liban, because this kind of behaviour had resulted in their expulsion.

Since the Liban story is used by the Burji as a discursive medium for reflecting upon and shaping their own identity and inter-ethnic relations, it is worth looking also at how the Konso and the Boran are represented and characterized there. Before doing this, I would like to make some general remarks about oral traditions.

Oral traditions are used as discursive media, but this does not imply that their meaning is fixed and invariant. On the one hand actual problems are reflected in the light of the past; on the other hand, the past is reinterpreted and sometimes reshaped in the light of the present. In short, the tradition A does not possess 'the' meaning B, but is given 'a' meaning by individuals or groups in a society. Anthropological interpretations of oral traditions increasingly take into account the dynamic interplay between text and context (see, for example, Vansina 1985; Finnegan 1992; Ben-Amos 1996). As Finnegan writes, 'irrespective of the particular approach taken up, the analysis of meaning almost inevitably leads outside the 'text proper' (insofar as this is something clearly distinct at all) to context, performance, control, and what people actually do' (Finnegan 1992: 185). Understanding and interpreting a given tradition thus depend on various contextual factors, such as situation of telling, social affiliation and the wider socio-political context (which is of special interest here). In order to understand the imaging of the Burji's relations with the Konso and the Boran in the Liban story, one has to establish connections between the contents of the tradition, indigenous comments on it and the socio-political context.

Representation of the Konso in the Liban Story

In all versions of the Liban tradition that I collected, the Konso are attributed negative traits. For example, the Konso are said to have split the communal bond in Liban by eating the sacrifical animal. Some Burji said the Konso's behaviour reflected a greed for meat, a greed that is still associated with them. Furthermore, the Konso tried to cast suspicion on the Burji by throwing the bones and intestines of the sacrificial animal in front of a Burji man's house. Due to their sacrilege the Konso are said to have been cursed to go naked and to have reddish teeth. When discussing the story in the Burji's homeland, additional, unflattering character traits were mentioned. The Konso were said to be clever tradesmen who liked to cheat customers. Unlike the

Burji, the Konso do not practise circumcision. For these reasons, and due to the incident in Liban, the Burji were said not to marry Konso.

The negative characterization of the Konso in the Liban story should baffle every anthropologist who knows about the close cultural and social ties between Konso and Burji that have grown up over a long period. Burji and Konso culture has so many traits in common that anthropologists have come to include the two in a category they have named 'Burji-Konso cluster'.[5] In the Ethiopian homelands of the Burji, there exist strong social, economic and political ties between the Burji and the Konso. Certain religious dignitaries of the Burji, holding the title of *ganni*, have family relations with some of the corresponding religious dignitaries in Konso, the *poqalla*. Burji temporarily or for generations have settled in Konso and vice versa. Burji may consult seers in Konso in order to know if war or epidemic diseases impend or to learn about the prospects of the future affecting the whole group or parts of it. Trading between Konso and Burji flourished in the past and still does up to now. The mutual relations are so close that there is an absolute ban on killing each other. One Burji remarked that, whereas the two branches of the Burji, the northern and the southern Burji, killed each other in the past, this would never happen between Burji and Konso. It is said that a Burji who has killed a Konso would contract a skin disease, come out in an itching rash or become sterile or his crops would not prosper. The Burji and Konso also form an alliance in times of war: whenever one of the two wages war or is attacked by a third party (for instance, by the Guji) they assist each other, 'like UN troops', as one of my informants put it.

How then is the negative image of the Konso in the Liban story compatible with the friendly relations between Burji and Konso? In order to find an answer to this puzzling question it is helpful to look at the indigenous comments on the story. When I discussed the Liban story with my informants in Burji, I asked how there could be good relations with the Konso when the latter had ultimately been responsible for the expulsion of the Burji from Liban. People answered that the incident had happened a long time ago and the Burji did not bear a grudge against the Konso for their behaviour. Others stated that the Konso had apologized. Some said that there was no problem between the Burji and the Konso because they were brothers (*fira*). Such statements, and this is a very important point here, were made by exactly the same persons who before had developed further the apparently negative image of the Konso in the Liban story. It follows that attributing negative traits to the Konso must be seen more as a form of mockery or teasing than that it should be taken seriously. Since relations between Konso and Burji are close, in the homelands at least, the Burji see no reason to stress this closeness as a central theme when narrating the Liban tradition. On the contrary, the familiarity with the Konso allows Burji to make fun of them. It is only if one establishes connections between the contents of the story, indigenous comments on it and the socio-political context

---

5. By the Burji-Konso cluster anthropologists denote a specific cultural type that has developed in the southern part of the Ethiopian Rift Valley. The name-giving Burji and Konso are just two representatives of this ethnographic cluster, under which around eighteen ethnic groups are subsumed (Amborn 1990: 26).

that one comes to understand that an unflattering imaging of the other can indicate one's closeness to him.

Let me continue with the relations between the Burji and the Konso in Marsabit in Kenya. The sociocultural, political and economic setting of this urban environment significantly differs from the one in the original homelands. It is to be expected, therefore, that this has an impact on how people interpret the Liban story.

Whereas in the Burji homelands agriculture is still predominant, many Burji in the diaspora communities in areas like Marsabit, Kenya, have middle-income occupations. There, they play a significant role in trading, especially long-distance and wholesale trading, and some have achieved considerable wealth. Marsabit is one of the largest and most important Burji centres outside the original homelands. In 1922, the British colonial administration at Marsabit invited a number of Burji peasant farmers from southern Ethiopia to take up farming near the little town because they needed crops for themselves, their clerks, police, local merchants, etc. At present there are about 5,000–6,000 Burji living in and around the urban centre of Marsabit (see Amborn and Kellner 1999: 9, fn. 2; Tablino 1999: 234). Before Kenyan independence, farming in Marsabit was the domain of the Burji. After independence, the government of Kenya allocated extensive tracts of the pastoralists' grazing lands around Marsabit to new farmers. Nowadays, the farmers are not only Burji but also pastoralists like Boran, Gabra and Rendille (Tablino 1999: 241).

The number of Konso living in Marsabit district is small. There are estimated to be several hundred families, of whom many are scattered. Most of them make their livelihoods by forging metal tools, weapons and ornaments and selling them to the surrounding pastoralist peoples. In Marsabit town a very small settlement of Konso craftsmen is found, consisting first and foremost of blacksmiths. In 1990, Hermann Amborn (personal communication) noticed in Marsabit town a severe pauperization among Konso craftsmen, a process that is still ongoing.

When I discussed the Liban story with Burji in Marsabit, the same negative clichés concerning the Konso were mentioned: the Konso did not practise circumcision; due to the incident with the sacrificial sheep in Liban, the Burji did not marry Konso; and so on. In Marsabit, however, these clichés had another, less joking flavour than in Burji. When I continued asking Burji in Marsabit what they thought of the Konso I was often confronted with derogatory comments. The Konso were just poor blacksmiths and made their living by forging iron, a 'despised' craft. It was also asserted that, apart from agriculture, Burji and Konso culture had nothing in common. The fact that anthropologists have placed Burji and Konso in the same cultural category, the 'Burji-Konso cluster', did not appear attractive to them. One Burji asked me, for example, 'Is there, anthropologically seen, really no way to leave the Konso out?'

The endeavour of Burji in Marsabit 'to leave the Konso out' is also mirrored in an article that was published in 1969 by Mude Dae, himself a Burji. Mude was brought up in Marsabit and later on started a political career, in which he became an ambassador of the Kenyan government. In his article, Mude stresses the cultural relations between the Burji and the Boran and claims cultural parallels with the Amhara and the Sidamo, but leaves the Konso completely out in this part. That is not all. He states: 'Legends apart, it is unlikely that the Konso could have come from Liban. There appears to be no real evidence of linguistic, racial or cultural affinity between the Konso and either these two tribes [the Burji and Boran] ... Furthermore, the Konso seem to belong to the negroid or nilotic racial grouping which inhabits parts of southern Sudan' (Mude 1969: 30).

Mude's statement being dated 1969 proves that the dissociation of the Burji in Marsabit from the Konso is not a new phenomenon. According to Amborn and my informants, relations between Konso and Burji were originally close, but changed after 'the *Shifta* War', which raged for several years from shortly after Independence (1963). '*Shifta*' is *Amharic* for 'bandit' or 'outlaw', and during this conflict many local northern Kenyan groups, particularly the (ethnically) Somali groups, fought for the areas they inhabited to be joined with Somalia. The conflict was bloody, with many killings, raids and massacres. Following the end of the *Shifta* War in 1968, the political and economic conditions in Marsabit changed, and this in turn changed the nature of Burji-Konso relations. In the parliamentary elections that took place in 1969 the Konso in Marsabit district voted for the Boran and not for the Burji MP candidate, as they continue to do today, an act the Burji continue to regard as a sort of betrayal.[6]

Economically, the relations between the Konso and the Burji changed following the reopening of the market in Marsabit in the 1970s, which had been closed during the *Shifta* War. Many Somali traders had been displaced, driven away, interned or imprisoned, due to the war, and Burji managed to take over much of their wholesale business. The Konso, for various reasons, were not able to make use of the new economic opportunities but continued making their livelihood with small-scale trade and forging iron. This new economic differentiation furthered the alienation that had been growing between Konso and Burji, and from that time the socio-economic gap between Burji and Konso in Marsabit has widened enormously. This may explain why the prospering Burji community in Marsabit does not like to be lumped together with these 'poor blacksmiths'. (It must be stressed, however, that the economic situation of the Konso in Ethiopia is much better.)[7]

---

6. In Marsabit, the Burji Woche Guyo recently met a Konso elder who 'acknowledged the oneness of the two' (Burji and Konso) and 'that it was politics that divided us' (correspondence, February 2002). Apart from the Konso behaviour at the polls, it is still to be clarified in what regard the Konso themselves actively fostered or intensified the alienation process between the two groups.

7. Ideologically, dissociating from the Konso in Marsabit has been eased by the idea of the 'despised craftsman'. Traditionally, craftsmen in south Ethiopia and in Burji are highly respected persons. Hermann Amborn (1990: 308–33) has shown in great detail that they are considered to be intimately associated with creation and entrusted with the performance of important ceremonies for the well-being of the community. With the Amharic colonization, however, the north Ethiopian idea of the despised craftsman with the evil eye has gained a foothold in south Ethiopia, especially in urban environments, and has not failed to influence people's minds. Apart from that the traditional economic role of craftsmen as producers and suppliers of crucial items, such as agricultural tools and clothes, has been severely undermined by the supply of industrial bulk goods. This is especially true for Burji craftsmen (*bijiri*), who have in addition to this – other than their colleagues in Konso – for various reasons even lost control of trading, which had formerly been one of their inherited domains.

One can also see the Burji attitude towards the Konso as developing from the Burji will to defend their dominant position as peasant farmers. Up to now the Burji are the most competent and skilled peasant farmers in Marsabit. The only group to compete with them in this field are the Konso, and therefore they must be kept out.

In summary, one can see how under certain conditions the teasing use of stereotypes can be transformed into ideological ammunition. Oral traditions give people core images to hand, which they interpret and develop further according to their needs. Another lesson to be learned from the example above is that one has to distinguish between the various regional fields and contexts when discussing inter-ethnic relations. One must not ask, for instance, what kind of relations exist between 'the' Burji and 'the' Konso but between Burji and Konso in different geographical areas.

Representation of the Boran in the Liban Story

In contrast to the Konso, the Boran 'come off well' in all versions I collected. The Boran expelled the Burji from Liban, but Burji show understanding for this as the Boran had been deceived by the Konso. Moreover, some Burji say a share of the blame had to be taken by the Burji themselves because they stubbornly refused to take part in the peace negotiations.[8]

In some versions, sociocultural parallels and familiarity with the Boran are stressed. As I have already mentioned, one narrator emphasized the similarity between the Burji and the Boran social order. Another narrator stated more than once that the Boran expelled the Burji from Liban by using clubs, not spears. One has to know that Boran do not spill Boran blood; a Boran who was sentenced to be executed was clubbed but without spilling blood (Baxter 1979: 70). Thus, the message which the narrator wants to put forward is that the Burji and the Boran were considered like family. It is also striking that in all Liban versions that I collected the sacrificial animal has been given a Boran name: *hoolaa faga*.

The question arises as to in what regard the apparently positive imaging of the sociocultural relations between the Burji and the Boran in the Liban story corresponds with the actual situation. It is noteworthy that in Ethiopia the Burji and the Boran were frequently at war with each other. Their territories overlapped, and in the search for grazing grounds Boran pastoralists often encroached on the lowlands of Burji. In connection with or in addition to the rivalry over resources, Burji and Boran undertook ritual killing raids against each other. In the first half of the twentieth century the pastoralist Guji, the arch enemies of the Burji, expanded their territory and moved in-between the areas of the Burji and Boran. Since then, clashes between Burji and Boran have been replaced by clashes between Burji and Guji.

Amhara slave raids in the first decades of the twentieth century led to a population drain in Burji (Amborn 1988: *passim*, 1995: 72). Many Burji sought refuge in Boranaland. The majority of them worked for the Boran as herdsmen or craftsmen.[9] In Marsabit symbiotic partnerships between a Burji cultivator and a

---

8. See note 4.
9. Woche Guyo, Marsabit, correspondence, September 2001.

Boran pastoralist were a frequent occurrence at least until 1980. Such partnerships could be established on the basis of clan or *gada* (generation-grade) relationships, but alliances also existed independently of these systems. When Boran in Marsabit took to farming, it was the Burji who helped them with their knowledge. With the booming of development agencies in the 1980s, however, Boran no longer had to depend on the Burji. Differences occurred, partially fuelled by Boran in the diaspora. In 1992–97, in the course of the tribal clashes in Marsabit, Burji were killed by Boran, and in 1999 I saw threatening letters from an obscure Boran youth association that had been sent to Burji households. As a result, the Burji are currently seeking alliances with Gabra and Rendille.[10]

In the Liban versions that were presented to me, the differences between Burji and Boran did not attract any response. This indicates clearly that the Burji have no interest in the deterioration of their relationship with the Boran. On the contrary, through using the Liban tradition as a point of reference the Burji devise a desirable future in which the previous good relations with the Boran are wished to be re-established.

Conclusion

The Burji reflect constantly upon their identity and inter-ethnic relations. In doing this, they refer to their orally transmitted history. The Liban migrational tradition, which is about the ethnogenesis of the Burji, is of special relevance here. It provides images and blueprints out of which Burji develop discursive lines along which they evaluate their present inter-ethnic relations and anticipate future possibilities in regard to this. In applying oral traditions to the actual reality, people make that reality meaningful and act on it, but at the same time reality has an impact on how people apply those given traditions. The Burji's discursive handling of the Liban story has revealed that when relations with another group (here the Boran) are strained or fragile but wished to be maintained, the inclusive potential of the Liban story is used by verbally conjuring up common grounds and the history of common 'good'; the differences are played down. When, as we have seen in the case of the Konso, inter-ethnic relations are substantially good and no big problem, it is not necessary to pick them out as a central theme. In contrast, negative stereotypes contained in the Liban story may be attributed to the other in the teasing sense, which, however, may be turned into ideological ammunition when relations with the other are worsening or cooling.

---

10.   See also Amborn's chapter on Burji in this volume.

Chapter 6

Mobility, Knowledge and Power: Craftsmen in the Borderland

Hermann Amborn

There is an indestructible trope that is perpetuated both by northern Ethiopians in the towns of the south, and even by European scholars. It is the idea that craftworkers anywhere in south-western Ethiopia are despised groups or castes. The artisans themselves have become aware of these defamatory notions, and have felt the effects of being considered in this light by outsiders.[1] In this chapter I shall discuss how they learned to cope with this problem, and the creative strategies they invented to reverse the power directed against them. I shall argue that this learning process had positive effects – largely unnoticed by the government – on their social situation and even on the well-being of their agricultural environment, and that this enabled them to create new forms of identity and integrity. As a striking example I have chosen to describe the formation of a craft organization with its centre in Konso.[2]

I first heard about this organization in 1974, during my ethnographic research in the Dullay-speaking area. I was at that time living in the compound of an uncle of the *poqolla* – i.e. the head of the leading *etanta* (agricultural) lineage of Gollango. One morning my host told me excitedly that he must leave immediately for Konso because an important craftsman (*hawd'o*) had died there. What surprised me was his state of excitement and the elaborate preparations he was making for the journey, which far exceeded those he normally made for funerals. Wearing clothes that identified him as a successful lion hunter, he finally set off, accompanied by several young men. Why was a craftsman being honoured in this way? After all, I had been told repeatedly in the *tej-bets* that no one would even drink with the *hawd'o*.

First I shall briefly consider certain economic conditions in this area; then, by showing the craft organization's development into an institution during the

---

1. For detailed discussion of the social, economic and religious position of southern Ethiopian craftsmen, see Amborn (1990). Pankhurst (1999) offers a critical discussion of the use of the term caste. See also note 7.

2. Cooperatives have been formed recently that sometimes show similarities with the northern Ethiopian *mahaber* and *ider* groups (on *mahaber* and similar groups, see Agedew Redie and Isabel Hinrichsen 2002). We can assume that there are also other cooperatives that we know nothing about. The networks of the Burji also originally go back to craftsmen, although these are of no importance today (see Chapter 4 in this volume).

twentieth century, I shall offer an interpretation of its significance, based on twenty-nine years of contact.[3]

Economic Area

On the southern edge of the southern Ethiopian highlands, two different economic systems border directly on each other: intensive permanent cultivation in the north and specialized pastoralism in the south. Regular exchange activities have doubtless been carried on along this borderline for several centuries, for on both sides it was possible to produce surpluses (Amborn 1990: 196–202). To bring goods for exchange, long feeder roads were and are necessary, for the border area alone cannot supply all needs. For mobile pastoralists like the Boran, travelling to the trading centres is no problem, for travelling is an integral part of their way of life. But this is not so in the case of the hill farmers. For them, trips to the market cannot exceed a certain length of time without putting their intensive permanent cultivation at risk. Constantly supplying distant markets is not a practicable occupation for them. Thus it was appropriate to delegate the job of trading with partners more than one day's journey away to a group of people not engaged in agriculture. Among the Konso and other south-western Ethiopian farmers, the division of work into agriculture and full-time crafts took place in the seventeenth century at the latest.[4]

Professional craftworkers (*xawd'a*)[5] always produce more than they need for their own use. Their products are normally sold in a market. In some cases, raw materials must also be procured from outside (e.g. metal for smiths). A smith, whether he obtains the metal himself or gets it through others, will thus be directly involved in trade. It is therefore reasonable that craftsmen should be considered as the most competent people in this field, which requires mobility.

A craftsman, who is not tied by agricultural work throughout the year, can change his place of residence without any great difficulty if he thinks the living conditions will be better in some other place. If he moves away, he often continues to supply his old customers, who have learned to appreciate his work. In the course of his life he can build up a whole network of relationships, which is passed on to his children and may be expanded by them. When a craftsman has a large number of customers, other craftsmen and even farmers give their products to him to sell and

---

3. In 1973/74 I lived with craftworkers in D'obase (Atano), Harso and Tuuro. I had close contacts with artisans in Gauwada, Gollango and Gorose, and made visits to Burji. In 1980/81 I lived in the artisan centres of D'iraša (mainly Yaaype and Kandikama) and Konso (Keera, with visits to D'ekatto and Tuuro). In 1981, I had interviews with Burji craftworkers in Marsabit; in 1984 I did a restudy in D'iraša (Kandikama and Kaalo) and Konso (several towns), and in 1991 in Badda Huri (north-western Kenya); in 1992 I had interviews in Arba Minch and Chencha, in 2000 in Konso (Turayte); in 2002/3 in Konso, as a member of the artisans' guild (*toola fuld'o*), I attended several of their meetings.
4. See Amborn, this volume Chap. 4, footnote 9.
5. Konso = *xawd'a* (a plurale tantum, although certain individuals form a singular); D'iraša = *hawd'a*; Dullay = *hawd'o* (Gawwada = *xawd'o*). Transcription symbols: d' = glottalized implosive voiced stop; x = voiceless velar fricative.

they ask him to obtain certain articles for them. When I was in the Dullay area, I observed several cases of craftsmen being pushed *nolens volens* into the role of trader in this way.

Mobile Artisans

With regard to the social position of craftsmen, it is significant that their outside contacts made them less fixed on their own culture than the farmers. It is true that this tended to make them suspect; yet these outside contacts were important for the community for they ensured the supply of necessary products such as salt. In times of poor harvest, survival of the community was guaranteed only through the *xawd'a* traders: they brought livestock from pastoralists, mainly from the Boran, or seeds from the inhabitants of the highlands in the north.

Another important factor here is the fact that artisans – no matter which ethnic group they come from – may not be killed.[6] They are considered sacrosanct and this makes it easier for them to work in foreign territories. Today, smiths and weavers are the most mobile professional craftsmen. Leather workers, and particularly potters, only rarely leave the area where they live. But the present-day decline in leather working has caused many practitioners of this craft to take to trading or to work as butchers in the marketplace. Let us take the movements of smiths as an example.

Most of their movements are within a radius of about 50 km. Unless they move to one of the smithing centres within this radius, smiths seldom stay away from home for longer than a few months. In addition to these smiths, there are some who seek a living in the pastoral societies, as is the case, for example, around Teltelle and Yavello. Even in northern Kenya (especially in Marsabit and in the Huri Hills) quite a number of smiths have settled. In fact, most smiths among the Boran today are of Konso descent. They provide them with metal objects for daily life and with ritual implements.

The life story of Usä, a smith from Kandigama near Gidole, is revealing. When he was in his teens he joined his brother, who had settled in Mašolle, to learn from him. Later he worked there as an independent smith. Following a dispute, he went to Yirgalem, where he stayed for two years. He then wanted to return to Yaaype, but on the way back he saw that Burji offered greater opportunities, and he settled there for the next two years. He then spent a year near Yabello and a year in Negele. After this came Kofolä in Sidamo, followed by Ginir in Bale. He then settled in Yirgalem again for five years, before returning to Kandigama, now a little over thirty years old. The people of Mašolle, where he had worked as a young man, asked him to come back to them. So, for about seven years, he worked in both Mašolle and Kandigama, staying in each for a few weeks at a time, until he finally felt he was too old for the constant changing. He then worked only in Kandigama, where the sales potential had increased.

The conditions described up to now cannot be considered apart from national politics and economic changes, as well as demographic factors. The conquest of

---

6. This was always true for D'iraša and the Dullay-speaking area, in eastern Konso, but not for craftworkers who participated in fighting.

southern Ethiopia towards the end of the nineteenth century brought about radical changes. Craftspeople were not spared by the general decrease in population. Many of them were enslaved or detached for compulsory labour. Entire crafts temporarily succumbed. In some areas, no artisans were left. At least two pottery centres disappeared from the map. It frequently occurred that craftsmen spontaneously moved to a different place at short notice, if they saw a chance of avoiding the tribute system by so doing. Later, in the early 1950s, it was leather working in particular that experienced an extreme decline after the Amhara authorities prohibited the wearing of leather clothing. Correspondingly this increased the weavers' profits. Weaving products are traded as far as Addis Ababa in the north and Marsabit in the south. Weaving is still a lucrative profession today, often exercised as a sideline occupation. For smithing, too, the new times brought not only disadvantages. Especially in the 1930s there was a high demand for smithing products, when the Italians constructed roads even in remote areas (Amborn 1990: 58–9, Freeman and Pankhurst 2001: 334). This also led to demographic changes among the craftworkers.

The ideology of the northern Ethiopians heavily affected socio-political life. Amhara and Tigre societies were characterized by a social ranking system ruled by clientele relationships, where craftworkers were found at the bottom of the hierarchical order. During imperial times, in parts of southern Ethiopia, this ideology was easily accepted, especially by those farmers who wanted to copy northern habits. The socio-political pressure they had to endure was transferred by them to the more powerless minority of the craftworkers.

This is not the case in those parts of southern Ethiopia where polycephalous societies counteracted the influence of the north.[7] This is in part connected with the spiritual power of craftsmen. In southern Ethiopian myths we find the motif of the craftsman in primeval times as an earthly demiurge, completing the process of creation. Thus man intervenes in creation more actively than in hierarchized 'literate' religions. Contemporary craftsmen are still considered to be gifted with this creative power. In such a world view a craftsman is a '*homo creator*' and not merely a subordinate hand worker (Amborn 1997b).

---

7. A century of northern Ethiopian influence has had the effect that, whenever ethnographers, and particularly those from northern Ethiopia, enquire about the social position of craftsmen, they first hear a repetition of this ideology. Dena Freeman, in Freeman and Pankhurst (2001), offers a categorization of craftworkers according to the agriculturalists' criteria: 'marginalised minority groups in southwest Ethiopia can be usefully categorised according to the way that they are stereotyped by the majority farmers' (p. 303). (This is an extremely doubtful method, which we shall discuss critically elsewhere.) According to her, all craftworkers can be classfied in four different categories of polluters (p. 304). In their book, the editors concern themselves with past research and add to it more recent research of varying quality. They base their arguments essentially on a great number of short studies. When dealing with such a sensitive and complex topic, the results of short-time studies must be regarded with caution, if only because of the recent debate on the aspects and problems of ethnographic representation.

Craft Associations and their Reorganization

The reaction of craftsmen to these changes was to form associations. Even before the conquest a strong feeling of togetherness had developed among artisans, who always formed a minority within the agricultural population. This found and finds visible expression in mutual attendance at funerals and weddings, for example. Although attendance at funeral rites is also binding for agriculturalists, this applies mainly in the case of friends or clan members. For artisans, regardless of clan membership or type of craft, it was and still is obligatory to attend all funerals of craftworkers within a certain area (which is greater than in the case of the agriculturalists). Because of their exogamic and endogamic rules, marital relationships also stretched beyond the immediate area of residence, but always stayed within the craft group.

Before the conquest by northern Ethiopia, craftsmen were usually organized on a local level. In Konso and Burji this meant in the towns, and within the craftworking centres in other regions of the Burji-Konso cluster. The form of organization seems to have been quite simple. A person from an important *xawd'a* lineage – i.e. one having important ceremonial duties – basically acts as an arbitrator.[8] The solidarity exercised on a local level, which – as *etanta* and *xawd'a* emphasize – always was, and still is, more marked than among the farmers, is considered as the basis of the present-day association. In it the line of tradition, creatively transformed, continues to be preserved.

By the beginning of the twentieth century, under the changed socio-economic conditions, organized associations were set up on a broader regional level. They can even be seen to include early elements of resistance. The activities of these associations included, for example, buying back artisans who had been enslaved by northern Ethiopians, by collecting money and negotiating with the slave holders. Negotiations regarding the exchange of *xawd'a* were also conducted between the various ethnic groups of the Burji-Konso cluster. This applied in particular to ritual experts, for without them important rituals could no longer be carried out (Amborn 1990: chap. IV, 2.2).

The first major, politically and economically effective supra-regional association was probably set up in Konso. It covered the three Konso regions of Takkad'd'i, Karatti and Tuuro (see Hallpike 1972). Here, around 1910, the craftsmen formed an alliance, when they realized that in this southern province far from the centre, despite coercive measures and tribute demands, the 'Amhara' were dependent on their craft products and on their cooperation.

In order to resist oppression, they took advantage of their exclusive slaughtering rights. They refused to supply the markets with meat, which mainly affected the *Ketema* (garrison town) population, less than the Konso themselves, since they continued to slaughter for them secretly. About ten years after the conquest, the Abyssinian soldiers, too, were no longer able to obtain livestock as easily as at the beginning. They had already stolen most of it, and what was left was kept hidden by

---

8. In 1974 it was still possible to reconstruct this historical period with reasonable certainty through interviews with the contact persons.

the local people and the neighbouring Boran as well. The action of the craftsmen was obviously successful, so that one region after another successively joined this association.[9]

In the 1940s the exchanging of goods in the area of the Burji-Konso cluster was nearly brought to a standstill when the land was terrorized by the so-called patriots (*fanno*) following the departure of the Italians (Amborn 1976: 154). It was due to the *xawd'a* associations with their complex network of contacts that exchanging was restarted on an inter-ethnic level.

In the 1950s there was an upswing in weaving. But this gave rise to new problems. The craftsmen were no longer able to meet the demand on their own and yet they were reluctant to let others take over this lucrative line of business. When some farmers secretly learned to weave, this caused tensions between the *etanta* and the *xawd'a*. Jensen (n.d.), who stayed in Konso in 1951 and 1955, reports that up to 20 per cent of the Konso were full-time weavers. A boycott on the part of the craftsmen against the farmers could not be kept up indefinitely – the demand for textiles was increasing too sharply for that. Yet alternation between the two socially defined categories of the *xawd'a* on the one hand and the *etanta* on the other was not possible without offending against social norms, even though many farmer-weavers used to call themselves *xawd'a* at the time. Instead, attempts were made to find methods for controlled integration of the new practitioners (Amborn 1990: 303). This also seemed necessary in view of developments in neighbouring Burji. There the weavers had split away from the professional craftsmen. Losing their most profitable occupation and their position of power, based on exclusive knowledge of methods of production and distribution, finally led to the professional craftworkers in Burji becoming an unimportant social group (see Amborn on Burji, this volume).

In the Burji-Konso cluster, after weaving had become effective for large parts of the population, a framework of reference based on descent groups no longer functioned adequately in the new situation. Some innovation was required. The nucleus and model for this were the local cooperatives, which up to this time had included only professional artisans. But now the principle of descent was abandoned, in order to make cooperation possible. And this gave rise to something completely new: *xawd'a* (craftsmen by birth) and full-time weavers among the *etanta* (agriculturalists by birth) now formed an institution together. In some ways this institution can be compared to the European guilds, even though the latter underwent very different socio-historical developments. But there was an important parallel: the new association regulated access to crafts, especially weaving, and entitlement to sell the products. *Xawd'a* of Konso belong by birth to the association, while *etanta* could become members of the association by meeting certain fixed criteria and performing a special ceremony. In our day they simply have to be confirmed. Thus professional knowledge continues to be at the disposal of a restricted group of people which, although now expanded, still has its own controls in place.

---

9. At first this involved only Konso-speaking areas. But there were dependencies in the Boran area as early as the 1920s. Gawwada (Dullay-speaking area) joined the association around 1930.

Discourses regarding this association show that the organization of the walled Konso towns *(paleeta)* may well have served as a model in the beginning.[10] They are subdivided into town wards. The town wards are corporative units with their own communal facilities and are inhabited by various clans. They are territorial units without kinship ties. Their members fulfil certain economic, social and religious tasks together. Also the different wards can rely on mutual aid. In some wards descendants of the first immigrant clans may have some privileges (see Hallpike 1972: 61). However, like the *xawd'a* in the association, they cannot decide alone, but are always dependent on the consensus of their cohabitants. In the new institution, too, territory fulfils an important symbolic function: here it is the *tiga fuld'o* (Plate 6.2), the house of the organization, under the roof of which persons from the two different role collectives, the *etanta* and the *xawd'a*, are united.[11] The association provides the means to integrate them, both at the social and at the productive levels.

Promoting Interactions across Ethnic Boundaries

It is known that in the early 1930s a craftsman called Kuyo Fuld'o, who was settled in Tuuro but also lived with one wife in Yabello (Boran), and who until then had functioned as *saara* (i.e. adviser) to the local organization, reorganized the existing association by applying new ideas and himself taking the top position as arbitrator. At the same time he strengthened the connections between Boran cattle herders and *xawd'a* from Konso. He is also said to have carried on successful negotiations with both the Italians and the English on the Kenyan border.

The reorganization of the social network, which had been destroyed by the *fanno* after the Italian retreat, is mainly due to the activity of Kuyo Fuld'o, who at the end of the 1940s and beginning of the 1950s, travelled from contact point to contact point setting up local committees, especially in the pastoral areas. He thus created the basis for the organizational form that still exists.

Although Kuyo Fuld'o's recognized qualities put him in a position of considerable power, he was not the 'chief' of the craftsmen. Thus, for example, middlemen were appointed by local craftsmen and not by him. Kuyo Fuld'o can certainly not be seen as 'elected' (in our sense) and yet there was a consensus among the craftsmen on his position. They considered him as their arbitrator and adviser, and as their representative in matters involving third parties.[12]

The success of this personality, who doubtless had charismatic characteristics, must have been impressive.[13] After moving to Purquta in central Konso, he was

---

10. Thus, for example, the function of the leading *xawd'a* is compared to that of an *aapa timpa* (the keeper of the sacred drum of the town and symbol of social order). On the *aapa timpa*, see Hallpike (1972: 47, 60) and Watson (1998: 219).
11. Of course not all *etanta*, but those of the *etanta* who adopted crafts.
12. The earliest mention known to me of Kuyo Fuld'o is in Jensen (n.d.) (fieldwork in Konso 1951 and 1955): 'Die chauda haben ihre eigenen Amtsträger, die auch chauda sind und für Streitigkeiten unter ihnen zuständig sind ... In Nordostkonso ist das Oberhaupt der chauda Fullo. Ihm unterstehen die chauda-Führer in den einzelnen Dörfern' (chap. B II.3).
13. Charismatic in the sense used by Max Weber (1976, chapt. III § 10), although he was far from being a prophet.

considered the *poqalla* of the *xawd'a* (Amborn 1990: 302), a title normally reserved for lineage leaders of the *etanta*. He also possessed important insignia of a *poqalla*, which are kept today (2003) in the *tiga fuld'o*. According to Kaala, one of the most important *poqalla* in Konso, as Kuyo Fuld'o was not a *poqalla* by birth, he was not entitled to them.

Because of the positive powers and intelligence that were attributed to him he was respected as an arbitrator by all craftsmen between Lake Chamo, Sagan and Dullay and beyond. Kuyo Fuld'o reconfirmed and strengthened the laws of *xawd'a* solidarity and transformed them from a local basis to the necessities of a broader network. He settled disputes involving members of the organization, which was later named *fuld'o* after his lineage, especially disputes arising between customers and members – which means that the farmers and pastoralists, and even fully acculturated Euro-Amhara merchants, also respected Kuyo Fuld'o's authority. In general, all foreign traders from different *Ketemas* in this area accepted his decisions, especially when outside the *Ketema,* even though they were in competition with the *xawd'a* traders (Amborn 1990: 302). Newly arrived artisans who were not accepted by the local craftsmen could turn to Kuyo Fuld'o, who worked out a solution for them together with other craftsmen.

Who was this famous Kuyo Fuld'o and where did he come from? The answer is multivocal and depends on the context in which the question arises. One may hear that he was originally an *etanta* but his mother was from the *xawd'a*; or that, many generations earlier, his forefathers had started to act like *xawd'a*; or that he was from a *xawd'a* lineage, but his mother was from Boran; or that four generations ago his ancestor was born out of a calabash, like many other *poqalla*, but with sorghum in one hand (like an *etanta*) and a hammer in the other; or that he was the last and only offspring of the oldest *xawd'a* lineage. The version told by craftsmen from Keera in Konso (1981) probably comes closest to the historical facts. According to this version, Kuyo Fuld'o's forefathers, who were members of a *xawd'a* lineage, came to Tuuro from Aylotta (possibly via Boran) four generations ago.[14] But some craftsmen in Keera and from D'ekatto also connected this lineage with the *etanta poqalla* Aylo of Aylotta. In this version the first Fuld'o was the twin brother (in another one he is the younger brother) of the one who became *etanta*. Here they were probably thinking of the myth according to which Aylo is the progenitor of all Konso clans, including the *xawd'a* (see Black and Otto 1973: 13).

Despite many attempts, I have not succeeded in finding any kind of authoritative genealogy of Kuyo Fuld'o. His son, or so I was told, was not interested in the matter, and even his third wife, whom I looked up in Yabello specially for this purpose, could tell me only the name of his father, which is unusual. Has his genealogy been deliberately suppressed? Is the direct connection to the earliest ancestor, without any intermediate links, meant to underline the special character of Kuyo Fuld'o, or is his line of descent kept hidden because there is something in it which does not fit the positive picture?

---

14. The Aylotta hills are in the most eastern part of the Konso-speaking area.

In these different stories, Kuyo Fuld'o's origins remain somewhat vague, consciously or unconsciously, but they all show that he was an integrative and charismatic figure, especially so in versions I heard in the 1970s. He was a focus for hopes of economic security, which were also shared by the *etanta*. The multiple versions of Kuyo Fuld'o's origins can also be seen as an indication of the role of mediator between *xawd'a* and *etanta* that was attributed to him and the institution. They can be interpreted as signifying that the *fuld'o* association, and Kuyo Fuld'o's prominent position in it, should be understood as something relatively new, and socially not yet firmly integrated in the cognitive system of the Konso. The stories point to an as yet incomplete social process of search for identity and attribution of identity.

One reason for the success of this organization is the fact that, although it occasionally openly opposed the government, it was still able to work virtually unobserved by the government until recently.[15] On the other hand, it could be sure of the support of the local people. Elsewhere I have discussed in detail the importance of the *xawd'a* for materially and ritually ensuring productivity in the Konso system of intensive agriculture (Amborn 1990: chaps II, V, 5). Here I shall only emphasize that in Konso outside contacts were made almost exclusively through craftsmen right up to recent times, for the Konso hill farmers are highly sedentary. It is traditionally considered improper to leave home frequently or for several weeks at a time. The *xawd'a* traders were the ones who imported new products, made contacts with other ethnic groups and introduced new ideas from outside. The state administration of the imperial era, however, was ignorant. Since according to their ideology craftworkers were socially insignificant, they failed to notice that an economically and politically strong and bold organization was growing up before their eyes.

The year 1974, when Kuyo Fuld'o died, coincided with the Ethiopian revolution. For the *fuld'o* organization this was one of the most critical phases. The new government took steps to put an end to discrimination against craftsmen and to promote handicrafts, but at the same time it was firmly against all existing 'traditional' institutions and created new associations that craftsmen were expected to join. In this critical situation, they missed their keen leader. The continued existence of the organization's network became threatened. However, after decades of experience in dealing with state authorities, the craftworkers of Konso succeeded in preserving their institution and even in reorganizing and expanding it, in accordance with the ideas of Kuyo Fuld'o. Through passive resistance they were able to dissociate themselves step by step from the forms of cooperation forced upon them.[16]

There were also changes within the crafts themselves. While weaving steadily increased, iron working and pottery have stagnated during the past thirty years (despite a slight increase in the population). Although industrial goods are offered for

15. Although it was not an underground organization.
16. My observations and information about cooperation among weavers and leather workers come mainly from Gidole (1981 and 2003). The *xawd'a* offered passive resistance as well as they could. There was no contact with the *fuld'o* organization.

sale, the people still prefer local products. Since the wearing of leather clothing was prohibited in imperial times, leather working has steadily declined; in D'iraša (Gidole), for example, the manufacture of leather sandals was discontinued about two years ago. Many former leather workers, who, like all craftsmen, had the traditional right of slaughtering, are now working as butchers in the markets. This occupation assures them a good income, due to the growing population in the *Ketemas*. As the number of animals kept for meat production by the peasant population is insufficient to meet the demand, cattle trading with pastoralists in the surrounding areas is becoming increasingly important. As a result, some former craftsmen have taken to cattle trading as their sole occupation. Thus, since the mid-1980s, there has been an internal division of work among those craftsmen specializing as butchers and traders. But there are still craftsmen who exercise their traditional right of trading and slaughtering in addition to their craft in the proper sense. Overall, however, there has been a shift of emphasis from production to trade. This went hand in hand with the opening up of the *xawd'a* group, which was formerly more closed. The prosperity of many *xawd'a* and the security offered by the guild made many *etanta* (farmers) want to join the *xawd'a* as a defined social group. While only about twenty years ago certain rituals had to be performed before *etanta* could be admitted as members, and even then they were treated only as 'learners' and not as full *xawd'a* (Amborn 1990: 300), today any person who joins the guild and accepts its rules is considered a *xawd'a*, if he wants to be.[17]

There was also a shift in the social definition of the *xawd'a* role collective. While only a few decades ago intermarriage was avoided both by the farmers and by the craftworkers, today such marriages are frequent (see also Watson and Lakew Regassa 2001: 264). However, even today only *xawd'a* by birth can become members of the decision-making committees.

These changes necessarily involved some reforms in the organization of the guild. It is said that Kuyo Fuld'o himself introduced the decisive reform on his sickbed. He, whose claim to leadership was based on the genealogical principal (he was a lineage head), handed over his position not to his son, who was still under age, or to a member of his extended family but to chosen dignitaries within the guild who were not even members of his own clan. They were to act in his name in future.[18] Kuyo Fuld'o obviously recognized that a new and more flexible form of organization had become necessary. Indeed, he had already begun to develop a polycephalous structure during his lifetime. He acknowledged that leadership based on the lineage

17. In the *tiga fuld'o* six trades are listed on a wall. The 'carpenters' have been added just recently. These are mainly construction workers who build tin-roofed houses. Some of them are *xawd'a* by birth, but the majority are *etanta* and craftworkers from other parts. In certain circumstances there is also the possibility of a loose association with the guild. For example, if a single *etanta* woman who occasionally sells alcoholic drinks in markets close to her hometown seeks the protection of the guild, this does not automatically make her a *xawd'a*.

18. 'Fuld'o was formerly a family name, but it is now applied to all people [who are associated with it]. Just as the council is called *toola fuld'o* and acts in its name. But the Fuld'o family does not exist [any more]' (Aapa Funnota Armanne Kayre on 14 January 2003).

principle would hinder the establishment of an association whose members are recruited from different descent groups. In addition to the older principles of organization modelled on those of the Konso towns, new forms were created that bore more resemblance to those of the mobile Boran. Particularly in spatial terms, these organizational changes permitted more flexible reactions.

We have already mentioned that there were economic exchange activities with the pastoralists. This gave rise to a paradoxical situation for Konso society: the *xawd'a* had close and often friendly relations with their Boran exchange partners, with whom marriage relations were arranged, while the *etanta* retained a rather distant or hostile attitude. Although this last has been increasingly shifted to the Guji (maybe as a result of the *xawd'a* relations), figures representing slain Boran are regularly found in the Konso groups of figures (*waka*) erected in honour of the dead (Jensen 1936: fig. 140).

The Boran are dependent on the Konso *xawd'a* and their products; this mutuality is sanctioned by the fact that the ceremonial headdress, the *halaša*, of the ritual dignitaries and the clothing worn during their installation must be made by Konso *xawd'a*, and that *xawd'a* must be present during the installation ceremony. Economic contacts between the Konso *xawd'a* and the Boran were accompanied by lively cultural exchanges (see below for details of the Boran committees).

A Flexible and Efficient System

Let us return to the central Konso area. The centre of the organization is the *tiga fuld'o* in Purqutta. The supra-regional committee meets in this house, which was built about ten years ago on the site of Kuyo Fuld'o's Konso compound, to discuss matters concerning the whole guild network (Plates 6.1, 6.2). The committee consists of a limited number of elected representatives. This committee, which is referred to, significantly enough, by the Amharic loanword *migir*, has formal authority, but it is extended by the addition of regional representatives whenever there is an issue affecting their region on the agenda. In addition, the meetings are open to all members who wish to attend.

The structure of the guild (*toola fuld'o*) is as follows: First, there are the local groups of the towns (*paleeta*), or, where there are no towns, of the districts. They look after the local affairs of the artisans and the regional markets, caring in particular for the maintenance of peace in the market. The organizational form of the committees is based essentially on pre-colonial structures. In Konso the nineteen local committees consist of the elder of a *xawd'a* lineage, supported by one or more elders, depending on the number of local *xawd'a*, and elected representatives of the *xela* (the most recently initiated generation group within the *gada* system[19]). If the local *xawd'a* are also active in the supra-regional network, they elect a *funnota* and, if necessary, a *melamitta* and *k'inaca* (see below).

Then there is a supra-regional committee, which is responsible 'for the road', as it is expressed. It is responsible for the trade routes and for the functioning of the

---

19. On the Konso *gada* system, see Hallpike (1972: chap. VI).

Plate 6.1 Members of the *toola fuld'o* in front of their assembly hall in Konso (photo: H. Amborn)

whole network within and outside Konso (see Map 4.1).[20] But in particular it maintains connections with the *fuld'o* branches outside Konso. This committee has a dual structure, with an eastern and a western section. Each of the two sections is composed of an *aapa funnota* (also called *k'ayeeta*) and his representative, the *melamitta*, and their assistants, the *k'inaca*. The latter act as their messengers. In addition, there may be between one and seven *xela* (also referred to by the Boran term *d'iirota*).[21] The two *aapa funnota* have the same obligations and rights, but the one from Karatti, since he lives in the central Konso area, is regarded as being the elder brother of the other *aapa funnota*. This means that his word may carry more weight than that of the 'younger brother'. Proceedings that only concern one section

---

20. Branches of the *toola fuld'o* outside Konso (most of them written on a wall in the *tiga fuld'o*): (1) regions within the Burji-Konso cluster (but outside Konso itself): Kusuma-Gato, Kaalo/D'iraša (with many sub-branches in the present-day D'iraša Special Wereda), Gawwada; accredited persons: Gollango, Burji; (2) north of the Burji-Konso cluster: Arba Minch,* Soddo/Wolayta, Boditi/Wolayta; accredited persons: Cenca (Gamo), Shashamane,* Auwasa,* Addis Ababa*; (3) west of the Burji-Konso cluster: Woito, K'ay Afer/Banna, Dimaka/Hamar, Bašada, Jinka (regional capital), Tabya/Hor (Arbore), Turmi/Hamar; accredited person: Omo-Rate/Dasanec; and (4) south and east of the Burji-Konso cluster: Teltele, Yabello (with sub-branches), Mega, Arero, Dubuluk, Hidilolo, Moyale (all Boran); accredited persons: Hagere Mariam, Yirga Caffe/Gideo. * = towns with a population of above 50,000.
21. *K'ayeeta* is used to refer to the institution of the *funnota*, but the word can also be used as his title. *K'ayeeta*, *melamitta* (also *melamica*) and *k'inaca* are not Konso words. *K'ay-* is perhaps from Oromo in the meaning 'to sit in front'; *melamitta* is probably derived from Amharic/Oromo *mäla*, 'solving a problem'. In Burji *milamme* can mean 'mediator between different groups' (*milammeši lukka god'ad'a* means 'the *milamme* makes relationships') (personal communication from Alexander Kellner).

Plate 6.2 Assembly hall (*tiga fuld'o*) of the *toola fuld'o* in Konso (photo: H. Amborn)

may take place in the house of the *funnota* (the *tiga funnota*) instead of in the *tiga fuld'o*. If the *funnota* cannot be present, the *melamitta* acts on his behalf. This principle is followed down the scale.

An *aapa funnota* is a dignitary with responsibility for half of the whole network, and his election must be very carefully prepared. The preparations may take several years. A candidate must possess diplomatic and rhetorical skill and be of irreproachable conduct. This also applies to his wife. She is included because she must be capable of making decisions in the house of the *funnota* in her husband's absence.

On a regional level each branch is responsible for itself. While the organizational structure is largely identical in all Konso towns, that of the branches outside Konso varies considerably. Yabello, in Boran territory, has sub-branches reaching as far as Arero, and follows the organizational principle of the *tiga fuld'o* in Konso. Here the committees consist only of *xawd'a* from Konso. In D'iraša, to the north of Konso, there is a *funnota* (who is not a Konso, but with affinal ties to Konso), but the rest of the guild structure is modelled on the relations between craftsmen that formerly existed in D'iraša. In Hor (Arbore), Konso persons play a decisive role in the guild, and also have close links with the local bond friendship networks and the local *ik'ub* groups. Decisions within the *fuld'o* team are made according to Konso and local rules (Tadesse Wolde 2002: 48–49). In Jinka, the organization is exclusively in the hands of locally resident traders who are not Konso.[22]

---

22. The situation in Burji is similar. Here, too, the guild formed an *ik'ub* at the same time, and money was paid into it together. However, the *ik'ub* was closed around 2001 after discrepancies were noticed. On *ik'ub* and similar organizations, see Agedew Redie and I. Hinrichsen (2002).

During the whole of its existence, the attitude of the organization to the state authorities has constantly undergone change, alternating between 'subversive' and 'cooperative'. Kuyo Fuld'o's successors in the committees obviously learned from his great diplomatic skill. He organized resistance at certain times, but at other times he was praised (so it is said) by Emperor Haile Selassie for providing safe trading links. And, as already mentioned, he is also said to have carried on successful negotiations with the Italians. Tadesse Wolde, who made a study of the *fuld'o* branch in Hor, comes to similar conclusions to mine, namely that the *fuld'o* 'went underground when governments were tough on them and became potent when government became weak and lost the control in the region' (Tadesse Wolde 2002: 50).

During my stays in Konso in 2000 and 2002/3, the relationship with the administration – as far as I could tell from my observations – was friendly, but tense and watchful. About six years ago, the *tiga fuld'o* had been renovated on the occasion of the annual celebration of the ruling party. It was painted in the national colours. Amharic inscriptions listed the different branches and gave information about the fields of activity of the members of the guild. Party members were invited to the opening ceremony, in order to emphasize willingness to cooperate. In return, the *toola fuld'o* was given a small office in Karati *Ketema*, the administrative town of the Konso Special Wereda.

The administration confirmed by contract that the internal decisions of the guild would be respected. But at the same time the administration laid down who was to function as chairman, vice-chairman and secretaries. The *toola fuld'o* reacted to this without any fuss. The persons named function as mediators between the guild and the administration, but continue to be normal members without any executive powers.

Solidarity and Capabilities

What are the aims of the guild and how does it set about achieving them? One aim is undoubtedly to assure production and distribution of the products made by artisans, including the procuring of raw materials and of cattle for meat and hides. Originally this involved preserving the market peace, and ensuring that craftsmen received a fair price for their work.[23] Cooperation with non-craftsmen was important here and, although the farmers were and still are dependent on craft products, it put craftsmen in a precarious position because they were in the minority. In order to strengthen their position in relation to others, as well as their own unity, they set up moral rules for themselves. These mainly concerned the obligation to provide mutual aid when need arose. This does not only concern business matters. The primary aim of the guild is to maintain internal peace and harmony among the *xawd'a*. Only

---

23. Some decades ago the tasks included the placing of orders for goods, such as leather and woven goods, which were traded in large quantities outside the Burji-Konso cluster. These discussions are insignificant today. They date from a time when it was important to control the production of the farmer-weavers. Today, the procurement of orders is left to the initiative of the individual.

when these are guaranteed can well-being and prosperity be established. The oft-repeated guiding principle applies here:

Atan   apaxaara   kod'anna,   atan an-neeqa   d'iina
We, who are with the good things adopted, we, who the bad things avoid
Olli   orsanneka   kallata pisa   kod'ame.
Together   we discuss   how our life together   we shape.
(Aapa Funnota Armanne Kayre)[24]

The desire for social peace and harmony comes to the fore during discussions in the assemblies, and the set formal hierarchy hardly matters any more. Anyone who can present his arguments in a good speech, enriched with examples from earlier proceedings, has authority. If the elders speak first without any formal preliminaries, while the young ones only speak after requesting to be allowed to speak, this is because the young ones are still gathering experience and need to learn the art of speaking through listening. And this is also the reason why the elders use many examples and proverbs. These are not used in a stereotyped manner, but as potential for reflection, in order to grapple with the changeable present through the use of traditional motifs.

No one has the means or the power to impose his own will. When a verdict is reached, all parties involved must agree. The aim is always to reach a consensus. If it is a legal case, the accused person must admit his fault if he is guilty and agree to the nature of the punishment. The wrongdoer is then cautioned and advised. The final act is a religious ceremony, through which the disturbed sense of community is restored. The offender is then free of any stigma. The case is then considered concluded. It would not occur to anyone to take the case up again at another level.

A good example is a case that was heard in April 2003 in the *tiga funnota*. Two young men, both from the same town, had quarrelled with each other. Instead of settling the dispute in their hometown, they continued to quarrel in public in other places, and, when they finally went to the *tiga funnota*, and found that the *funnota* was not there, they continued to quarrel there, too. The point at issue in the subsequent meeting was not the reason for their quarrel, but the impropriety of quarrelling in public and even in the *tiga funnota* itself. By doing this they had breached the social harmony of the whole group. The proceedings lasted almost two hours. The speakers repeatedly quoted examples from the past of right and wrong behaviour, which they underlined with proverbs. The aim of these speeches, spoken vehemently and with practised rhetoric, was not only to demonstrate the impropriety of the young men's behaviour and thus to find a measure for the severity of the punishment, but also to instruct and advise them by showing them examples. At the end of this instruction, they begged the *aapa funnota* and each person present to pardon them. They accepted the well-debated punishment (which was the same for both of them). The apology was accepted, and the event was sealed by the elders

---

24. we-1pl of-good make-imperfective:1pl; we-1pl of-bad avoid together discuss-with-and living always it-is-made.
I thank Maarten Mous for the word-by-word translation.

giving their blessing to them and to all those present, thus confirming that unity had been restored.

Anyone who participates in such *xawd'a* assemblies will see many parallels with the Boran assemblies, as described and analysed by Bassi (1999). For the *xawd'a* these Boran assemblies offer a guide; the *xawd'a* appreciate the experience of the Boran in respect of supra-regional committees. The Konso *xawd'a* sometimes imitate their manner of conducting discussions through example. This finds expression, for instance, in the frequent use of Boran legal proverbs, in the quoting of whole passages in Boran or in the so-called 'slaughter the coffee' ceremony (*punnita qali*), an important component of the final ritual following the successful conclusion of proceedings. This was directly adopted from the Boran. Even the terms used to designate dignitaries in the guild are partly derived from Boran terms; and the insignia of the *funnota*, the whip and the ceremonial staff, are of Boran origin (see Haberland 1963: plates 25: 7 and 9b). In the latter case, this is a new development, for these emblems were obviously not used by Kuyo Fuld'o, or at least they are not among the insignia of Kuyo Fuld'o preserved in the *tiga fuld'o*.

These examples show that Konso *xawd'a* find their identity not within Konso society alone; rather, they have creatively expanded it through careful adaptation of other cultural elements. This not only helps them to feel comfortable in other cultural environments; with their rituals and above all with the new insignia (to the Boran insignia they have added others of their own invention), they have succeeded in creating a counterbalance in Konso society to the ritual position of power of the *etanta* lineage heads (*poqalla*).

Among the cases treated by the *toola fuld'o*, a large place is occupied by those concerning the repayment of debts. The *xawd'a* would not be able to fulfil their orders and carry out most of their activities without credit. This ranges from money for buying cotton with which to weave a man's shirt, to credit for the purchase of a herd of animals for the meat market before an important holiday. With a very few exceptions, every *xawd'a* is a debtor and creditor at the same time within his community. Failure to repay is therefore not just a private matter between two people, but always affects a whole network. The whole system of production, acquisition and sale can only work in this non-capitalist society if all those involved can mutually rely on each other.[25]

In order to explain why it is necessary to deal out punishments, one of the committee members gave the following example:

> If the *xawd'a* from our town are robbed by another *xawd'a* from a different place, our *xawd'a* could say, let us now go and rob them. In order to prevent

---

25. In an assembly held in April 2003, a debtor who had already been ordered to pay in previous proceedings, was in such an extreme predicament that he faced the threat of expulsion from the guild. During a break in the proceedings which lasted only four minutes, half a dozen of his acquaintances got together and hastily arranged credits, possible pledges, advance payments and methods of repayment, which then enabled them to produce the cash which the accused was required to deposit immediately with the *aapa funnota*.

this, we act in accordance with our cultural law. We order the wrongdoer to come to the *toola fuld'o*. There he is told: 'You have cheated and robbed someone. Now wherever you go the people might think, because you are a bad man they can take money by force from you, or from anyone in your village because they think everyone there is a robber.' Therefore he must be punished in *fuld'o*'s house.

However, if an accused person refuses to acknowledge the rules and decisions of the *toola fuld'o*, he faces the threat of expulsion from the community. All branches are immediately informed of any such expulsion, which means that the outcast can no longer work in the whole area of influence of the *toola fuld'o*, and nothing will be sold to him. Even bus drivers will refuse to take him to another place. Usually, after some time, an outcast will come back asking to be forgiven. But once there was a case where forgiveness was impossible: there was a *xawd'a* lineage that was involved in the slave trade before the Italian occupation. It was cursed and expelled from the *fuld'o* association. The power of the *fuld'o* organization is revealed by the fact that this decision amounted to expulsion not only from the *xawd'a* but from Konso society as a whole. It is believed that most members of this lineage have died by now, but some of them are said to be still living far beyond Yabello.

Of course, some wrongdoers deny their guilt, if only to avoid the fines, which may be extremely high. The method used to establish the truth is simple: the accused is repeatedly requested to tell the truth, but this process may last many days if his words are not believed. Yet sometimes in the *tiga fuld'o* the accused may be placed between the house posts from Kuyo Fuld'o's former house, which have been installed there. This was explained to me as follows by the *aapa funnota*: 'We say to him: "Say the truth." So it became swearing by itself for him.' If someone tells lies in the *tiga fuld'o*, he must expect terrible punishment. The *tiga fuld'o* with its old house posts obviously symbolizes the smithy; and the tradition still holds that in the smithy the truth must be told.

At any rate, the *toola fuld'o* is so successful in settling disputes that even northern Ethiopians in Konso prefer to go to it rather than to the courts. Furthermore, this reputation promoted the incorporation of rich and influential northern Ethiopian traders, which in turn had another positive effect: these traders, although they are themselves dependent on the organization in the south, have brought it additional prestige (see also Watson and Lakew Regassa 2001: 263). The high standing of the *toola fuld'o* is also connected with the fact that one of their main tasks is to secure the paths of communication between the individual branches. And indeed the organization is currently more effective than the police in regions such as the border area in the extreme south-west of Ethiopia. Security also means ensuring that sufficient water and food are available, during cattle drives for example. In addition, care must be taken that women and children are not cheated, that those in the marketplace who are physically and financially weaker are allowed to do their business first and that helpers receive a fair wage. If a rich man makes an unusually high profit, he must give a share of it to the poorer members.

The effectiveness of the *toola fuld'o* was demonstrated once again in the spring of 2003. In the Lower Omo, a conflict-ridden area, as we can learn in this volume, *xawd'a* from Konso had bought about thirty head of cattle in the Omo-Rate market. Very shortly after the purchase, they all went missing. In the subsequent difficult debates, which lasted for three months, it turned out that administrative officers, foreign merchants and truck drivers who did not belong to the *toola fuld'o*, as well as cattle raiders from among the pastoralists in the surrounding areas, were involved in the case. Although Omo-Rate was very far away and had no *fuld'o* committee, thirty cows were brought to the *xawd'a* in Konso after the successful conclusion of the proceedings.

According to statements by guild members, the success of the *toola fuld'o* and people's trust in it are the reason why more and more regions have joined the organization. And this has contributed to the fact that the *toola fuld'o* is probably the most influential indigenous organization in the area south of Arba Minch.

Enriched Identity

In conclusion we can say that the *xawd'a* community in Konso has undergone several changes in the past hundred years, before reaching its present status. In a first phase, before and some years after the conquest by Menilek's armies, there was strong social cohesion within the *xawd'a* on the local level. This is the period that always serves as the point of reference, even today. At that time the *xawd'a* had no identity problems. They were a necessary and therefore fully integrated role collective for society as a whole. The rules of community life were bundled together in the second phase of Kuyo Fuld'o and creatively developed and expanded through the careful and skilful inclusion of foreign cultural elements. In a society not bound by written rules, this succeeded through flexible adaptation to orally handed-down traditions; in such memory cultures this adaptation can be seen as an integral part of reflective processes constantly recomposed in accordance with specific situations. At the same time, their openness to foreign cultural elements enabled the artisan community to gain access to new fields of activity outside Konsoland. By intensifying their communal bonds and at the same time opening themselves to influences from outside, the artisans were able to avoid being marginalized and pushed to the periphery of society under social, economic and political pressures from the north. Kuyo Fuld'o played a decisive role in this development; it is therefore not for nothing that his lineage name, Fuld'o, stands today for the whole institution.

Within the institution itself the most fundamental changes have taken place during the past few decades during the transition to the third phase following the death of Kuyo Fuld'o. Where everything was once focused on an intelligent and even charismatic leader, there is now a council that functions through mutual agreement. In the historical situation, i.e. under the pressure of the imperial administration and its ideology of the craftsman as an unclean pariah, the keen and charismatic leader was just the right person to organize the artisans.

Under the protection of the topos that the dominant society was taken in by, and at the same time with support from their own community, the craftsmen were able to develop a micro-subculture through the *fuld'o* institution. They were able not only to preserve their endangered autonomy, but also to extend it. The external pressure did not fail to produce a counter-reaction within the institution, increasing its awareness of its own power: not of a power aiming at subordination, but – to put it in the words of Hannah Arendt – '[A] power [that] corresponds to the human ability not just to act but to act in concert.'[26] In other words, a creative power has the ability to make collective decisions and solve conflicts. This new organizational power was a hidden, but essential element in the cultural resources, which was now used to produce structure and action. Under these conditions it did not matter any more whether there was a single charismatic and powerful leader. The person was unimportant, for cooperation resulted in a collective increase in power. This was doubtless in interaction with economic success. Such a situation could be established only because it was the Konso community, including the clientele outside Konso, which first constituted this power because of their recognition. The interplay of these factors resulted in a renewed and increased sense of dignity for the *xawd'a*, based on their culturally rooted symbolic capital (Bourdieu 1993: 216). The confidence growing out of this new dignity gave rise to a new identity on a broader basis which was created in the process of formation of the association.

(Manuscript completed in August 2003.)

---

26. Arendt (1970: 44). In her book Arendt put stress on the distinction between power, strength, force, authority and violence (1970: 43–56).

Part III
Land, Identification and the State in Ethiopia

Chapter 7

'We Have Been Sold': Competing with the State and Dealing with Others

Tadesse Wolde Gossa

This chapter provides an overview of the relations between the people of the Gamo highlands and their land, and between the Gamo people and the Ethiopian state. It explores the impact of the state in the period of conquest and incorporation, and the state's subsequent conduct regarding its subjects, particularly in relation to the nature and impact of land alienation.

The Gamo people number about a million, and they live in the mountains and surrounding lowlands of the Gamo Gofa Highlands in southern Ethiopia. They understand and explain the initial stage of the process of losing their land at the beginning of the twentieth century as 'selling'. This experience has been so profound that they also relate it to themselves, saying that they have 'been sold'. According to this expression, they are left without power and have become 'effeminate', attributing maleness to those who conquered them. As maleness is attributed to their conquerors, it is assumed that the productive capacity of the land, and of women, is notionally taken over by the conquerors; to be defeated is taken as a loss of their manhood, together with their capacity for fatherhood and ownership of any property. Although they express their feelings in this manner, the way they have been coping with life under the conquering state, the Italian colonial state and the post-colonial state until 1991, demonstrates that they have been doing everything within their means to regain fatherhood and ownership over their land. The way they have dealt with the state over the years has also varied, and does not show a complete loss of 'maleness', as understood by the Gamo and their neighbours. They have dealt with the state in ways that range from confrontation; to using the state system to complain against the evil practices of provincial officials; to competing with the state for resource use; to migration as an adaptation to the growing pressure on land and state dominance. Throughout this period, they were told that they were Ethiopians, but, in their dealings with the state, they have often witnessed double standards; they can be said to have been considered more as subjects in the colonial sense than as citizens.

The argument of this chapter is simple. The interaction of the state and its conquered subjects in the south-west over the last one hundred years indicates a regular pattern in the behaviour of the state. The state's appetite for land, for forced conscripts or for labour and its demand for material or financial contributions in addition to taxes have been insatiable. This behaviour, and particularly the way the state has been responsible for the alienation of rural land, has led to growing discontent and to confrontational relations between groups and the state. To support my argument I take the case of the Gamo, and explain how the state's dealing with them over the years has affected their views of others and has also shaped the ways in which they identify themselves.[1]

I start by giving a range of examples that illustrate Gamo notions of land and the way they link to religion, understandings of power and what it means to be Gamo and non-Gamo. I then give a review of Gamo experiences during the conquest and the Italian colonial and the post-war and revolutionary state.

Land and Sacrifice

When I visited him in the summer of 1995, Shagire, a Gamo man of about sixty-seven years of age, suggested that I slaughter a ram in gratitude to the route that brought me safe to his house. I did as he advised. I took the ram to the centre of the path passing next to his house, which runs from Chencha town towards Pango, to Zad'a, to Ele, to Gughe and to various other destinations. While I slaughtered it, I said the following prayer according to his instructions:

Ogezo, ogezo
Zaruman gatstidayso
Heko nes

Oh path, the path
You that have brought me here safe
There you are (This is for you)

Paths connect people with other people and places. They take one to markets and back. People use them to travel back and forth. Diseases come and go using these same paths. War comes along paths and leaves along paths. That is why, when we come home after our travels, especially from distant and unfamiliar places, we have to make a sacrifice and express our gratitude to the paths we have taken, for the peace and well-being with which we were bestowed during the journey; we wish that it may always be so. Although the Ethio-Italian war was the last war to be experienced in these parts, other wars have required many thousands of conscripts to be taken from this region. Many have not returned. The sacrifice for the *oge* path has become more relevant than ever before, as new forces and developments have come that affect the lives of people.

---

1. The data on which I base this chapter are mainly qualitative. They were gathered through interviews made between May and July 2000, and are also based on literature compiled by geographers in the 1960s and on personal experiences made during three years of work on a dictionary project in the 1980s, as well as on working and living in Arba Minch. They also draw on my experiences from belonging to the community and speaking the language fluently. I have not used many comparative data, although my own study among the Hor, some work among the Konso and some literature on the neighbouring peoples have been quite useful.

On another occasion, after travelling to Doko to visit him, I slaughtered a cock in Shagire's compound for dinner. Shagire does not eat chicken as many Gamo refuse to eat birds. The cock screamed while dying. A while later, Shagire suddenly fell ill. He held me responsible for his illness and said, angrily, that I should not have killed the chicken on land that is his domain, and not mine.

In his land in Doko, only Shagire and his elder brother, if he had one, or his lineage head, are entitled to slaughter. Neither his eldest son, Wale, nor his younger brother, Solomone, can slaughter while he is alive. Only the death of the father, or his approval, entitles the eldest son to slaughter for food or sacrifice. A sacrifice can only be performed by eligibility acquired through seniority on ancestral land.

On still another occasion in 1995, I invited Shagire, Ade Indala and a few other friends to visit me in Hor country, where I was doing fieldwork. I told them that they could stay with me for a few days and get to know some of my Hor friends, and that I would treat them with Hor *dad'di* honey wine and Hor-style roast lamb or goat, *wadde*. They accepted my invitation, but Shagire noted that they were not allowed to kill animals outside their land of Gamo. The spilling of animal blood must take place on one's own land. The slaughter of an animal is either for *Ts'os* or for *Tsalae*. The former is an all-powerful deity of the heavens, and the latter of the earth. Gamo use such opportunities to read the intestines of animals and to understand current situations.

On another occasion I told this group of friends about my interest in making a feast to receive the Gamo title of *halaqa*. They told me I was eligible to take the title, but that I could only be given the title in my native settlement, known as '*dere*', and on land that is my father's or that I have inherited or own in some way. The title can only be given to me by my *dere* settlement and on land where the ancestral centre-post is erected.

These experiences gave me insights into the links between a Gamo person, *Ts'os*, *Tsalae* and land. The link with the supernatural can be effective only if mediated through sacrifice performed on land to which one is connected. This legitimizes the person not only as owner of the land, but also as the proper authority to communicate with the supernatural. From other lands, others can communicate with their own gods.

I was not able to understand the relationship between the people and their land clearly until I compared my experiences of Gamo with those of the pastoral Hor of Lake Stephanie in southern Ethiopia. I always thought that the Hor view of their land was of religious proportion: they loved their land and they would not perform any major ritual outside their country. After their incorporation into the Ethiopian state, they were often in exile among the Dassanetch, the Boran, the Tsamako and the Hamar. But in order to perform major rituals, such as the transfer of power of generations, they had to wait until they could return to 'their country'. Even today in Hor, major rituals such as weddings must be performed in the village to which one is affiliated as a Hor. I mention this comparative case simply to indicate that the Gamo case is not unique in southern Ethiopia.

Land and Livelihoods

In daily conversations, Shagire and his friends swear in the name of assembly places or in the name of mountain pastures where they herded animals as young children. Sometimes the term truth, *tuma*, is used to swear by. At other times, when pressed to prove that what is stated is true, names of fathers are invoked. Herding boys and girls use the names of mountain pastures they herd their sheep on in the same way. People avoid using God's name *Ts'os* as a swear word; only urbanized people in places such as Arba Minch and Chencha occasionally use *Ts'os, maramo* or the names of Christian saints as swear words.

The cases shown above indicate that the relations between land and persons are intimate. The *zuma* mountain pasture commands near-religious respect and every person in the mountains has a personal link to it. Every young person's childhood is spent in these pastures with animals and unforgettable memories of annual rituals and feasts. To a large extent, barley, wheat and enset cultivation depend on manure that is sourced from animals grazed on the mountain pasture (Forster 1969: 435).

The Gamo in the mountains have two homes: a main home in the valley and a secondary home, known as *mots*, close to the *zuma* mountain pastures. The main home in the valley is considered to be senior; it is a ritual centre and the main base for the family. It is usually inherited by the eldest male and has ritual significance over the land and the houses of younger brothers. This is the house with the centre-post of the ancestors. The family, sons and daughters and spouses of sons live in and around this house. Milk cows, ploughing oxen, riding horses and sometimes draught horses and riding mules are kept there. The main barn and the main manure piles are also there. Enset groves surround the yard, and bamboo groves abound on the lowest and wettest part of the land. Other trees mainly of hard wood (*koso, lolashe, anka*, etc.) and *borto* trees, whose foliage is used for fertilizing the fields and as bedding spread for animals, are grown not far from the main house. *Zode* thorny berry plants are usually planted around plots to create barriers against animals destroying the crops. All major events of the family take place here in the main compound. Usually up to eight houses of different sizes and functions are built within it. Married sons are provided with a small house, allotted to them at marriage, and they stay there until the death of the father. After this event, land is divided between brothers, and the compound of the primary house ceases to be the place where male siblings live together. Only ritual relations are maintained, with the younger brothers and sisters (who marry out) becoming dependent on the inheritor of the main house.

The secondary homes, or *mots*, on the mountain pastures, are surrounded by bamboo fences. They may also be surrounded by trees and enset groves. Koso trees, which have good leaves for providing bedding, are planted. Large ferns which are used, when dried, both for animal bedding and for cushioning in beds. The *mots* contains one or two thatched houses and a sheep shed built on top of a pit, covered with a sparsely woven bamboo mat. This allows sheep droppings to be collected in a pit, which are transported and applied to the fields as manure by women. Nearly all sheep and cows are kept in this *mots* house (Jackson 1970: 18). Secondary houses are cared for by children, both boys and girls, or by newly-wed couples.

There are also secondary *mots* houses at the foot of the mountains in the lowlands. These rarely house sheep because the surrounding land does not require manuring. Instead they are surrounded by fields for growing lowland crops and for keeping cattle. Complementarity exists between the main houses and the *mots*, and between the ecological areas. The relations between the pastures and the fields are fundamental to the production of subsistence and to the maintenance of society. *Mots* (noun), of which the verb is *mod-*, is the act of establishing a settlement and fields or cattle camps in various ecological zones to reap the resources in different agro-ecological zones, including the use of pastures on mountains as well as at lower altitudes. Gamo land tenure allows mountain dwelling *dere* community members to have land and common pastures and forests. A *dere* community such as Dorze, for example, has a mountain pasture and *mots* in the mountains, their main homes in the mid higher grounds and their lowland *mots* in the escarpments and lower grounds – thus having a long strip of land that extends from mountain top to the lower lands.

The mountain pastures are taken care of by hereditary officials known as *Maka*, who are custodians of the mountain pastures and, by implication, of the fields. *Maka* organize yearly feasts for young shepherds in their houses. Each family with herding children contributes some grain to give to the *Maka*, which he uses for a feast in the festive season of *Masqala* at the end of September. The *Maka* erect plaited bamboo poles in the midst of the upland pastures, signifying a prohibition of grazing during the rainy season leading up to the festivities of *Masqala*. Mountain grass, which has been left to grow over the rainy months of July and August, is opened at *Masqala* for animals to graze unattended.[2] This is a festive season when people socialize in the home and in assembly places, markets and pastures. People working in the fields and herding children take a rest from their arduous tasks. Migrant weavers also return from central Ethiopia to attend the *Masqala* ritual, bringing with them cash and gifts.

Apart from the two types of house, there are multiple named and hierarchically ordered sacred places. These include sacred groves, assembly places, marketplaces, places where oaths are taken (marking a peace deal between neighbouring groups), hierarchically ordered cemeteries of craftsmen and citizens, sacrificial sites, cross-paths, swamps, streams and *zuma* pastures. Some of the sacred places also serve as assembly places and places for celebrating festivities and holding funerary dances. They can be described as the heart of the landscape. Each of these is named in a way that maps the intimate relations between people and their land. Each public place also has a hereditary custodian official such as the *Maka*, who protects this public

2. Jackson (1970) also mentions this practice. He reports that, during their research in 1968, the *Maka* in Ch'ento were no longer performing these functions. My evidence shows that they are still working, however. I have recently met some of the *Maka* who still have control over the use of most of the upland pastures and they are supported by their respective *dere*. In one of the *dere* I visited, a prophetic movement is under way to protect the pastures further. The threat now appears to be from NGOs and from evangelical organizations that embolden their members or beneficiaries to use pasture-land and sacred grove resources as proof of their devotion to Christianity rather than tradition. This is in addition to the state's interference in the use of pastures for state forests.

space and makes sure it serves its purpose in the reproduction of society and culture. A number of taboos also maintain and reproduce tenure relations and relations of social hierarchy and construct the common good. For example, any sexual play in the fields is a taboo and has to be cleansed instantly to maintain the fertility of people and the land. Similarly, it is a severe offence to move boundary markers, to plough pastures or to use any of the common spaces in a way that contradicts their role. Selling any field close to pastures other than to a family member is forbidden, unless it is to the *dere* community, who will add it to the pasture. This has produced the local saying that 'pasture-land gives birth to more pasture', an arrangement that is upheld by the *Maka* official. Just as every primary and secondary home and fields are under the ownership of a male head of a household, and a group of such households is under the ritual authority of a senior figure of the clan, his primary house and the land on which it is located, so each person's access to common places is mediated through the *Maka*, whose authority in these and other religious matters is acknowledged by the community concerned.

The lower valleys and lands on the slopes in both the east and west of the Gamo mountain range are communally owned by each *dere* community adjacent to it. The escarpment to the east and the lands along the Lakes Abaya and Chamo are the focus of this chapter. The lower lands along the two lakes used to be, at least four decades ago, the *mots* secondary home and field equivalents for those who lived on the escarpments. Here each *dere* owned a large chunk of land roughly between its respective settlement and the lakes. *Dere* such as Ganta (south-east Gamo) owned the land further across the lakes, covering the area that is now the domain of the Arba Minch National Park, Nechsar. Their land reached Gad'a Bonke, and as far as the hot springs at the foot of Mt. Yero.

These lowland secondary homes were used as a base for growing lowland crops.[3] The Gamo are known as weaving people all over Ethiopia and their cotton produce is still sent across the lakes (Abaya and Chamo) to markets in Sidama, Ghergeda country (Gujiland), Wolaita and Dawro, across the River Omo and to Addis Ababa. The clothes they produce are worn for religious and other life-marking rituals. Lowland crops, such as maize, sweet potatoes, bananas, coffee leaves, *haytse tukke*, from the escarpment settlements and other fruits, are also produced and exchanged for mountain products, such as enset, barley, wheat and pulses. In addition, meat and dairy produce from the lowlands were taken to mountain markets and sold for domestic consumption.

The lowland zone provided and still provides other supplies useful for domestic animals and for markets in the mountains. *Ado* is excavated from the shores of the lakes and is transported to mountain markets, where it is sold as a salt lick, providing essential minerals for mountain cattle and small stock. People from the mountains come there with large numbers of packhorses and transport many hundreds of sacks

---

3. In the lowlands the main agricultural areas were limited around the Rivers Wajifo, Dudane, Baso, Hare, Kulafo, Sile and Elgo and other streams for irrigation. Local cotton and some other crops were planted during the rainy season and did not require much irrigation, such as maize and banana.

for domestic use and for sale in the markets in Bodo, Pango, Zozo, Ezo, Zad'a Giyassa, Geretse and Balta Ch'osha and in other markets on the western escarpment. All *dere* are entitled to this. They were also entitled to *shanka* hunting in the lowlands, as these grounds along the lakes were rich in game and wild beasts (Hodson 1927). Young men from various *dere* came to the lowlands to hunt as part of a ritual for ending the period of seclusion after circumcision. There were no *dere* restrictions against the use of these facilities. These grounds were also used for obtaining specific lowland woods used for the construction of houses and for making ritual staves for dignitaries and shafts for spears. The conditions in the lowlands were also favourable to bee-keeping.

The ritual leaders of the *dere* that command authority and claim ownership over this lowland terrain are highly regarded for their power. They are known for containing the evil in the lowlands and for protecting animals and the lives of people on the mountain *dere* from predatory animals that prevail in the hot lowlands. The mountain people give such *dere* particular respect because of the way they deal with their enemies, particularly in pushing back military threats that used to come from the country of the Gergheda.

The western Gamo lowlands extend north to south from the River Gogora to the eastern escarpment of the Maale country. Borodda, Ch'illashe, Qogo, Shoch'ora, Kullo, Ts'ela Quch'a (west), Manana, Halaha, Dara, Malo, Shella, Anko, Kamba and Zala are some of the *dere* in this configuration. This region is well known in all of Gamo and beyond for its butter and aromatic milk and soft cheese. Dairy produce used to be transported from these *dere* to the Selam Ber market on the Sawla-Addis road and from there to Arba Minch, Addis Ababa and Soddo markets. The wealth of dairy produce can be explained by the availability of good pasture and the excellent animal husbandry of the people.

The State and Land in Gamo

The conquest state

Evidence from ancient manuscripts and ecclesiastical objects in some ancient monasteries indicate that the Gamo were a section of a larger pre-Ethiopian politico-religious entity. Bahrey, the author of the *History of Galla* is said to have lived in the Monastery of Mariam in Birbira.[4] In a map drawn by Merid Wolde Aregay (1971), showing the probable extent of the Ethiopian empire at the beginning of the sixteenth century, Gamo is indicated as part of the political unit.[5] He also cites Paez, who writes that Emperor Susenyos told him that Bahr Gamo was the southern limit of his empire (Merid Wolde Aregay 1971: 21).

---

4. Bahrey (1954: 155); Tafla (1987).
5. In a later discussion he makes a distinction between Gamu and Gamo and Bahr Gamo and Suf Gamo without realizing that the former was a label used by the then Ethiopian authorities rather than by the Gamo themselves (Merid Wolde Aregay 1971). Merid also does not substantiate his claim that the Kuera and Gamo, currently residing on the east and west of Lake Abaya, formerly resided further north between lakes Zway and Abyata.

In recent history, however, the Gamo were incorporated into the Ethiopian empire by Menelik's forces, led by Woldeghiorghis. The conquering army reached Gamo immediately after Wolaita was subdued in January 1895. The war of conquest, and what it cost its victims in terms of life and property, was horrifying. Looting of property, massive slave raiding and indiscriminate killing were features of the conquest.[6] There were widespread disease and hunger as a consequence. Only some of the early experiences of Wolaita, Maji and Kaffa have been documented by travellers, chroniclers and European military advisers or, more recently, by historians and anthropologists of the south-west. Their accounts fit well with the oral traditions of the descendants of the victims of conquest.

At the initial stage of conquest, *dere* settlement residents were shared out between the conquering officers and soldiers, according to rank and title.[7] The conquering officers fought among themselves for the Gamo farmers. This system, known as *gabbar*, was a form of tenure in which the conquered people's produce and labour were extracted for the maintenance of the conquering soldiers and their leaders based in garrison towns. As farming communities were made into *gabbar*, they and their families became mere slaves of the conquerors. Valuable agricultural land, pasture-land and sacred groves were taken away for settlements and for setting up churches, and land around the garrisons was also taken and given to the soldiers (Jackson et al. 1969: 8). Families and their produce were assigned to soldier families to provide them with supplies and labour. Assembly places were de-sanctified. The Gamo use the term '*nuna bayzida*' – 'they sold us' – to explain the events in which land was captured and people coerced. The soldiers in the mountain garrison towns of Ezzo, Koddo, Chencha, Gulta, Bazza, Balta Qara and Bussa extracted produce and

---

6. Bulatovich (2000). Richard Seltzer, the translator of Bulatovich's text, also writes about the regions that refused to submit to Menelik: 'he turned [them] over to his most talented commanders, whom he let have the opportunity to conquer them and "feed off" them. However, once these regions had been completely destroyed by war, they could not supply provisions to all the troops that had conquered them, which gave rise to the conquest of neighbouring lands which were still free' (Bulatovitch 2000: 218).

7. Olmstead (1997) reports Chimate's opinion on the *gabbar* experience of the *dere* of which she was *balabbat*:

   The Amharas from the north who settled in the area were given local families as *gabbar*. A low-ranking soldier would be given one family, a high-ranking administrator a hundred. Each family had to give four days of labour per week to the person it served. Every Sunday, people brought wood, water and grass. A yearly tax was paid, in cash, and every month a thaler's worth of butter had to be given. For food, each family gave the Amhara barley, wheat, and chickpeas. If a family said, 'We won't give', they were imprisoned. If there was not enough, a household member was taken as a slave. (Olmstead 1997: 33)

   Olmstead adds that King Mijola was given fifty families from his own people. While Larebo's focus seems to be on the ideology of *gabbar* and its human face, Chimate's quotation emphasizes the lived experience (Larebo 1994: 35–37).

labour from the population.[8] The late Seyoum Desta, himself the son of a former *gabbar* owner, once told me that the *gabbar* system provided comfort not only for the soldiers, the clergy and the politicians of the time, but also for their riding horses and mules. These were kept and fed inside *gabbar* houses, 'like a privileged new mother in seclusion who had just given birth'.[9]

The leaders of these *dere*, known as *kao*, were descendants of *dere* founders and were assigned to positions of local power due to their acceptance of the new authority. Fearing a terrible supernatural consequence of taking responsibility for the new system, most *kao*s passed the new positions of *balabbat* to their younger brothers, so that they could maintain their traditional role. The exercise of power in Gamo required complementarity between the power of hereditary ritual chiefs and an elected assembly of elders, who ran daily affairs of each *dere* community by discussion, persuasion and consensus. Each member of the assembly was at one time or another chosen to serve a term as messenger of the land, after which they were allowed to join the ranks of *der adde*, 'fathers of the land', who can then sit on assemblies for the rest of their lives.

The incorporation process was extremely hard to bear for the Gamo. At one stage, the custodians of some of the ancient churches in the Gamo mountains went, dressed in their ceremonial clothes, carrying the tabot arks and some of their ancient

---

8. The conquering soldiers had formed settlements in many garrison towns. One such town was Koddo, which was a pasture ground of Zada prior to its selection as a garrison town and occupied by the conquering soldiers of Menelik in the mid-1890s. Four hundred soldiers and officers were based here under Fitawrari Tekle Mariam. It was heavily fortified, and the surrounding peoples of Gamo from Zad'a, Lisha, Kullo, Hlalaha and Selo Dinke were made *gabbar* to the officers and soldiers (Asane 2000, personal communication). Pankhurst (1985: 199) mentions the type of fortification of the then Gamo town of Ezo, which is situated about seven kilometres north-east of Koddo:

    Chencha was reinforced by two nearby fortified posts, Ezo and Goddo [Koddo], both built in the aftermath of Menelik's occupation. Ezo, which contained a number of houses and gardens separated by bamboo partitions, was remarkably well fortified. It had a triple line of defence. On the outside it had a trench, two metres wide and deep, which was half filled with water by the rains; in the middle, two metres further in, a circular wall of stones, two metres thick and a metre and a half high, surmounted by a barbed palisade; and on the inside another trench. Entry could be effected only in three places where the wall was pierced by three gates with corresponding bridges. Goddo [Koddo] which lay to the east [south-west] of Ezo had a similar system of defence.

    All garrison towns found during the incorporation had a number of *dere* that were assigned to them as *gabbar* land.

9. The land tenure in the north and south varies considerably to the extent that to list the types simply becomes confusing. But an excerpt from Bulatovich's longer explanation of the distribution of property may simplify the matter: 'Galla lands together with their population belong to the emperor by right of conquest. All Galla [refers to the conquered peoples] are considered obliged to pay rent, and at the same time the process is beginning, which took place in Russia at the time of Boris Godunov – the process of turning people into serfs' (Bulatovich 2000: 84).

manuscripts, and made an appeal for peace. The Gamo hoped that by doing this they could avert what they saw as an imminent catastrophe. Their hopes died when the custodians were taken as prisoners to Addis Ababa, where they were held in Menelik's imperial palace. Many of their artefacts were also looted at this time by the military leaders. The custodians were only able to return home after an epidemic ravaged the population of Addis Ababa.

It is commonly thought that it was the fertile and coffee-growing areas of Ethiopia that attracted the most attention of the conquerors, and that areas such as Gamo mountain land received much less interest because it required frequent manuring and hard work to be productive. However, the loss of land in Gamo in this period should not be overlooked: in addition to the land on which garrison towns had been built, fields surrounding these towns were taken. Sacred groves used for religious sacrifices were taken over by the Orthodox Church, which established their church buildings on them. In later developments, sacred assembly grounds and marketplaces were targeted as sites for constructing schools and police stations. The Gamo were on no occasion consulted, nor have they ever received any compensation for their loss of land.

Haile Selassie's attempt to modernize the state machinery, particularly his effort to end slavery and the *gult gabbar* tenure system,[10] did not have much effect, as governors and their soldiers continued their practices illicitly, and in most cases were opposed to reform. The *gabbar* relations continued to be an effective means of extracting wealth, although the Emperor assigned his men to various areas to supervise the activities of his officials.

Confrontation between the conquering soldiers and the population around garrison towns only grew tenser when the news of the coming of the Italians was heard. The Gamo had not been planning revenge against their conquerors but this was the agenda of the Italians. When the Italian conquest became certain, the Gamo started to refuse to pay tributes. They said they would wait and see whether the Italians would also demand tribute, as it would be unwise or impossible to pay and please two masters at the same time. The *gabbar* masters took this as a betrayal and declared war on the Gamo. A war was fought under the leadership of Dejazmach Woldemariam (husband of Askala Jote, an Oromo woman of noble background from Wollaga). Forty-two men and three women were brutally executed and accused of inciting a rebellion against *gabbar* owners. Many houses were burnt and much property was confiscated. Many women and children were taken as slaves.[11]

---

10. 'Gult simply means "grant", but popular opinion put the *gult*-holder into a distinct and powerful class. In fact, *gult* indicated a territorial unit of administration where the state renounced part or all of its fiscal rights in favour of a *gult*-ruler' (Larebo 1994: 33–34). While this may theoretically be correct at the centre, the experience of the Gamo and the way they talk about this tenure and the relation between the *gult*-holder and *gabbar* tenants is different.
11. Asane (2000) 'Life in Zad'a', personal communication (recorded interview), and Seyoum Desta (1990), personal communication (recorded interview).

The Italian period

After taking control of much of the country, the Italians put Gamo under the *Governo della Galla-Sidama*. The Italian colonial authorities banned *gabbar* practices and, at the same time, they encouraged the killing of Amhara. The former decree was welcomed, but the latter was completely rejected owing to an earlier oath that had been taken between elders of the two communities. The unnecessary spilling of blood is discouraged in Gamo and is considered a severe offence, resulting in dire consequences for both the performer and the land in general.

Although the *gabbar* system was abandoned, the Italians reintroduced something similar. They forced the Gamo to labour in road construction and to provide Italians with construction wood, labour, milk products, poultry and grass for their animals. The labour required included the carrying of ammunition and food supplies to distant places such as Bulchi. The labour conditions were extremely harsh and people were treated brutally. Young people were conscripted into armed contingents to help subdue groups further away.[12] Work on the farms was disrupted and life grew miserable. Life was much worse than it had been in the *gabbar* days, and Gamo people in general became embittered. Soldiers of Hamasien origin from Eritrea and from Somalia worked alongside the Italians, and were present in great numbers and based in the Italian camp in the sacred assembly place of Gembella in Dokko. There were also partisan groups, who would punish anyone who they thought might be collaborating with the Italians.[13]

The Italians also tried to foster inter-ethnic enmity and killing. This was largely a failure in Gamo, however, except in one location between Wobbara and Dita. Here Italian soldiers forced some Gamo to inform on where partisans were based and, as a result, many hundreds were caught and machine-gunned by Italian soldiers. The place of massacre is locally known as 'Amaray Wurosa'. Gamo have left this spot fallow ever since. The Italian project of taking land and settling armed units to farm, which Larebo (1994) discusses with regard to other parts of Ethiopia, did not materialize here, as most of the Italians' time was taken up with the construction of roads to Gidole, to Gofa and Bako and to Kamba over the Gamo mountains and with fighting partisans. (For information about the experiences of the Konso and the Hor under the Italians, see Watson 1998 and Tadesse 1999.)

The post-war state

When the post-war government of Haile Selassie was established, the Italian administrative structure was rearranged so that there were twelve provinces (Ethiopian Mapping Authority 1988: 2.5). The new administrative unit was named

---

12. Prior to the coming of the Italians, young Gamo men had been forced to join various campaigns while at the same time fulfilling their *gabbar* obligations. In these campaigns they were used as porters as well as being made to fight alongside their masters.
13. Fitawrari Kanko of the Qogo *dere* ordered the hanging of a man from Chencha for informing Italians of the hiding place of some Amhara (see Chimate's accounts in Olmstead 1997). The Italians shot Masa, the king of the *dere* of Dita, for hiding seventy boxes that belonged to the Amhara in his house, and for allegedly telling his people not to kill Amhara.

Gamo Gofa and its capital was first established at Ghidolle, where the mainly Oromo-speaking soldiers of Menelik had a garrison. The capital was then moved to Chencha (also shown in the literature as Dincha), and this became the centre of the new Province of Gamo Gofa.

When the post-war administration of Emperor Haile Selassie launched its modernization programme, it built the first three schools in Borroda, Chencha and Gardulla (a different district), and started installing its state administration. *Gabbar* was officially replaced by tax and this marked the beginning of the salaried civil service. But the new officials saw their appointment in the province, or into the civil service, as a reward for serving the Emperor loyally during the war against Italian colonists. This attitude was coupled with the traditional central Ethiopian notion of power and government, in which the ruler was viewed as all-powerful. In the provinces, the rulers and their followers had unrestricted power over their subjects and property. They took control of all fertile settled lands, escarpment pastures and forest groves, and took lowland pastures as *qalad*[14] both for themselves and their followers. *Qalad* owners were entitled to collect tax from people who lived on the land and worked on it. In return they also paid tax for this entitlement.

Over the years until the Revolution of 1974, the Ethiopian government recognized communal land tenure arrangements in northern Ethiopia. In the south, however, only settlement areas and individually owned cultivated land were recognized; non-settled land and communally held land were declared state property, which could be disposed of at the will of the authorities, disregarding local forms of tenure. Even when the new tenure arrangements were introduced, it was clear to the provincial and higher government authorities that some forms of tenure coexisted with the state tenure system. The state tenure system, people were told, was superior to the local tenure system, as it 'facilitated development'. The Gamo, however, noted that this was not quite the case and, suspicious of the official tenure system, held theirs intact. Wherever their land fell under the *qalad* type of arrangement, however, they still had to pay annual contributions to the person the state tenure defined as its owner.

In the early 1950s, the development policy of the Ethiopian government focused on settlement projects involving farmers. The Ministry of Community Development and Social Affairs was the main institution through which, in 1953, the central government launched agricultural development projects in Awasa (former Sidamo), Bottor (former Showa) and Arba Minch. These projects aimed to develop state-supported plantation-type mechanized farms, which would assist selected families, each of whom were given five hectares in the project farm. The state would plough, provide water for irrigation and supply seeds, insecticide, machinery, technicians and

---

14. *Qalad*, as used in Gamo, is land given to people of central Ethiopian origin by the state during the imperial days for services given to the state. It may be bush land or may be cultivated land with holders. The *qalad* owner collects tax from users of the land. While the *qalad* owner was recognized as the owner by the state tenure, in Gamo tenure the person who tills it and whose ancestral land it is is acknowledged as the owner. *Qalad* owners knew this and did not evict farmers.

advisers. The farmers were expected to follow advice and to provide all the manual labour required on their plot. The government took responsibility for selling the produce from the farm; it deducted its own costs and paid the settler farmers the remainder. According to the model of the farm organization, the farmers would eventually develop into a commercial cooperative.

The Bottor farm failed and its employees were assigned to the Arba Minch and Awasa projects. These projects performed well according to their set objectives. A detailed discussion of the development of these farms is beyond the scope of this chapter; here I want to emphasize the impact that the Arba Minch farm had on wider developments in the region. The success of the Arba Minch Development Farm (AMDF) in Gamo attracted a great number of daily labourers from Gamo country, and from groups such as the Wolaita, Konso, Gidolle, Kore/Koera and Burji.

The then Governor of Gamo Gofa, Dejazmach Aimero Selassie Abebe, was so impressed by the development of the AMDF that he arranged with Addis Ababa for the capital of Gamo Gofa to be transferred from Chencha to Arba Minch.[15] When Arba Minch was established as the capital, the area of land taken from Gamo was well over 1600 km^2 or 160,000 hectares. Farmers who had cattle camps and fields of various crops were evicted from their lands in favour of the AMDF, the new town in the making and the Nechsar National Park, attached to the town. Only a handful of *qalad* landowners of central Ethiopian origin were singled out for compensation; no Gamo who owned the land were either consulted or given compensation. A few years later, in 1968, the Addis Ababa-Arba Minch road construction was completed, and, between Addis and Arba Minch, a thrice-weekly bus service and a twice-weekly air transport service began.[16] A ferry service was started across the two lakes that connected the province of Gamo Gofa with Sidamo. The establishment of such infrastructure and the favourable investment atmosphere that the government introduced made Arba Minch attractive, and the demand for land alongside the lakes in eastern Gamo rose. Thus, the fertile *mots* land of the lowlands, which was owned by various Gamo communities (Wajifo, Dudane, lower Donne, Ankobare, Oumo, Lante, Zayisse and Ganjule) from the River Hamasa in the north to the River Wozaqa in the south, was given as property to various categories of people including Europeans, rich and middle-class central Ethiopians and high-ranking civil servants with entitlements to state land (for their contribution during the Ethio-Italian War and for loyal service to the state). The last two were entitled to forty hectares of rural land. This was opposed by members of the Gamo communities of mainly Otchollo, Dorze, Chencha, Sul'a, Ezo and Dudane, some of whom were living in Addis Ababa as weavers.

The first two kinds of incomers, the Europeans and central Ethiopians, established plantations with state support. The central Ethiopian farmers were given loans that included a generous grace period, low taxation and tax-free fuel for agricultural purposes. The interest of the farmers was in developing cotton

---

15. The land taken for the town of Arba Minch belonged to Ganta farmers of Gamo. It was their *gad'a*, lowland *mots* and cotton field area. It was locally known as Doysa and Garo.
16. The airport was completed in 1964 (Jackson et al. 1969).

cultivation, as there was high demand for cotton to feed the national textile industry in Qaliti, Addis, Dire Dawa, Bahir Dar and Asmara. Arba Minch became one of the main cotton-producing areas.

Whereas the labour required for planting, watering, thinning and picking cotton was readily available from the surrounding peoples, the uncertainty about the reaction of the Gamo to the taking of their land created fear among the developers and government officials. Accordingly, a police training centre was established on the edge of the National Park, which produced ordinary and emergency police. Side by side with this, the imperial territorial army was strengthened in the area. The southern command of the territorial army was headed by Dejazmatch Tadesse Wolde, former governor of Konso and veteran of the Ethio-Italian war.[17] The problem with this unit, however, was that a central Ethiopian background was a prerequisite for membership. People with this background already in Gamo tended to live in the mountain garrison towns, but they were Gamo-ized and on extremely good terms with the Gamo. Moreover they remained in their mountain holdings out of fear of lowland malaria. Officials thus decided to bring in new settler farmers from Shoa, Wollo and Gondar. Settlements for these northerners were arranged in pockets of land taken from the Gamo for this purpose. The first settlers were recruited particularly from Merhabete and Menz in Shoa. The rationale on the surface was that the Gamo were 'lazy' and did not know how to farm and therefore these farmers would teach them how to make use of the land. The real reason, however, was to have armed contingents that did not depend on the state for a living but could crush any resistance to the plantations. For this purpose settlers were given land and were made to join the imperial territorial army, which entitled them to firearms. They were based in Wajifo, Shelle and Gatto. Although the planners of the plantations had expected danger, it did not, in the end, come from the expected source. Problems were to emerge later, not from the Gamo, but from the state itself.

New sites for settlement along the Arba Minch-Addis Ababa road were started at this time, at Shara, Lante, Channo and Wajifo. According to their plan, the Ministry of Community Development and Social Affairs was to provide a development worker for each settlement, and to build each settler family a house in a model village. Social services such as clinics, schools and veterinary clinics were part of the package. At first, the idea was welcomed by the Gamo farmers, who assumed that they, the former owners of the land, would be resettled in new, serviced and equitable arrangements. Later, however, it became clear that, in the villages of Lante and Channo, communities would not be able to keep their members together; outsiders were also being accommodated among them. Jobless people from Arba Minch, for example, were rounded up for settlement. The Gamo farmers objected to this vehemently, and the settling of others, particularly unknown outsiders, was aborted. All the same, some land was still kept aside following this incident. Pockets of land ranging from five to forty hectares were given to officials, to their families and to officials at the centre. When this became clear, the people of Otchollo and other

---

17. His two sons were successful cotton farmers on their plantation along the River Wozeqa.

Gamo communities mentioned above, who were then living in Addis Ababa as migrant weavers and parliamentarians, went to Emperor Haile Selassie's court and demanded an end to the land-grabbing. The Emperor, it was said, gave them a caretaker, promised a solution and assigned his personal chief of guards, General Assefa, to handle the matter. The General was a personal friend of the governor of the province. The next day, the General arranged for the rounding up of those who had appealed and put them in a police camp in Kolfe, whence they were transported in two trucks back to Gamo.

The effort of the Gamo to use legal and appropriate channels to reclaim their lands was frustrated, and their confidence in the system was thwarted. In the province itself, measures were taken to intimidate those who were suspected of opposing the taking of land. Tensions became heightened. In one incident an army colonel who was given forty hectares of forest land around the Arba Minch-Chencha junction cut down the acacia forest that belonged to Otchollo, and started making charcoal for markets in Addis Ababa. When Gamo residents went to talk to him, to advise him to stop chopping trees until the decisions of the Emperor were known, the men on the farm shot and wounded some of them, which ignited a series of demonstrations in Arba Minch. A second incident involved the ploughing of ancestral graves of the Zayse close to the Shelle-Gidolle junction between the Rivers Sille and Elgo. The YILGE plantation, which two central Ethiopians jointly owned, cleared the grove on top of the graves, and ploughed it for cotton, exposing skeletons that had been buried. This incident also caused great public anger and embarrassment to the provincial officials.

By 1973, most of the lowland cotton plantation was owned by one European, by high-ranking provincial officials, by settlers from central Ethiopia, by merchants from Arba Minch and by rich young farmers, such as the owners of YILGE. The remaining lowland *mots* land was owned by *qalad* owners, who were mainly descendants of Menelik's soldiers. Armed settlers from Shoa occupied three pockets of well-watered fertile land in Gatto, Shelle and Wajifo. During the 1973 famine in northern Ethiopia an additional group of settlers, mainly from the provinces of Gondar and Wollo, were brought and settled in north Gamo, in Manukka, in the western lowlands that belong to Borodda and close to Salam Berr (Selo Dinke), along the River Deme in the western Gamo lowlands. These settlers, like the earlier arrivals from Shoa, were fully supported by the state, and were, like all central Ethiopian groups listed above, given land that was the location of Gamo lowland secondary homes.

In the lowlands, the Gamo were now in strong competition both with the state and with the groups of people that the state had given their land to. As a consequence they also started planting cotton and cultivating land that they usually left fallow or as forest, out of fear that otherwise the state would appropriate it. This competition between the state, the Gamo and the new owners was so high that, in less than three years, nearly all the forest resources outside the National Park became depleted. This increased the need for waged labourers, not only for the state farms, where they constructed irrigation schemes and worked on the cotton, but also for the Gamo smallholders of Shelle, Lante and Ch'anno. The main source of labour was Konso men who had initially been sent as convicts to the central provincial prison at Arba Minch. There were also a large number of daily labourers from Wolaita and Kore.

In late 1973, as tension was building up between provincial officials and the Gamo over land, the central government was facing serious opposition from university students and mutinous soldiers in various barracks. When the situation in the lowlands deteriorated, the provincial administration armed civil servants of central Ethiopian origin against the Gamo who they suspected would exact revenge for the loss of their land. Of course, this was an overreaction to the tension, and represented an underestimation of the Gamo people's recognition of state power.

Until the mid-1970s there were no Gamo in positions of power in the provincial state bureaucracy. Some were employed in clerical positions in various provincial and district government departments. Schools had a good number of Gamo teachers. All local *dere* state appointments (such as *balabbats*) were held by local kings and assemblies.[18] Of these, two were women.

Throughout all these developments, people tried to maintain the complementary role of their mountain primary homes and lowland secondary homes. Many Otchollo and farmers of other *dere* went up and down the mountain every day after work in their fields in the lowlands out of fear of malaria. Festivities, major family and *dere* events and religious and economic interactions were carried out in the highland *dere*. However, with the settlements flourishing in the lowlands, in competition with the state and land-grabbers, the lowland homes gradually underwent a process of transformation into main homes. The establishment of clinics, schools and veterinary services and the widely admired effort of the state to eradicate malaria were conducive to enhancing this transformation.

As the revolution gathered force throughout the country, Arba Minch was ready for it. Those whose land had been alienated and those who worked for one *birr* a day (about ten UK pence) in cotton plantations swarmed into the town of Arba Minch. 'Land to the tiller', the popular slogan of Haile Selassie's students, had strong local currency. As is the case in such moments, both victims and land-grabbers, even middle-level civil servants who owned small plots, joined the public demonstrations. The provincial administration and their families and the large plantation owners left, seeking refuge in the centre of Ethiopia.

The Derg state

After the coming to power of the Derg in 1974, the March 1975 proclamation made rural land public and state property. This meant the end of land-grabbing in rural areas and, in a way, liberated peasants from their century-old servitude. At the initial stage it appeared that all lost land would be recovered, but it also heralded the emergence of a state monopoly on mechanized farming. While farmers were expecting the restitution of land that had been given to private plantation owners, the state made itself the legitimate inheritor of all such farms. All plantations in the Arba Minch area were made into state farms. There were unexpected results as well:

18. To understand the institution of *balabbat* in central Gamo see Olmstead (1997).

Gamo farmers who had employed labourers from neighbouring groups in the Lante, Chano and Shelle areas were seriously affected. Labourers employed in the fields could now be recognized as having rights to the land they tilled. A farmer who had previously owned five hectares and who had four labourers working for him was left with only 1.5 hectares of land. On the other hand, all Gamo farmers who lived and worked on *qalad* land were able to take back their land.

Eventually, new developments emerged. Loyal to the traditions of the Ethiopian state and regardless of its socialist ideology, the state continued to be excessively involved in the daily lives of the farmers. Like the conquest state and the pre-war, colonial and post-war Ethiopian states, the socialist state not only was involved in the alienation of land but began deciding the kinds of crops farmers should produce. It set quotas for farmers' associations (which replaced the *chiqa shum* office of the imperial structure) and the amount and type of grain farmers were obliged to sell to the state. Furthermore, it organized a corporation for purchasing grain for prices set by the state.

Paying no heed to traditional tenure, it targeted mountain pastures, escarpment common land and lowland common areas, and used these places to develop state forests. None of these lands can be considered to have been, at the time, 'free'. The largest mountain pasture, Sura, was owned by four *dere* (Doko, Lisha, Chencha and Ezo) and was made into a large state forest. This affected the fields/pasture ratio and created an imbalance in manure/field relations, which meant a considerable fall in crop yields and an increase in pressure on land. The pastures of Mazze, Harpa and Dayda (owned by Doqama and Zute) and Gircha (owned by Chencha, Doko and Ezo) and many other mountain pastures were partly or wholly made into state forests without the consent of the people. What is worse, according to the custodians of the pastures and the general farming public, was the fact that the exotic trees that were planted not only contributed to the fragmentation of the soil, and hence dryness, but also destroyed grass that would not normally have been affected had the trees planted been of indigenous varieties.

The interest of the state in land was a constant threat to farmers. The land reform decree drew an image of the state as a force that put an end to previous atrocious practices. The reality, however, was similar to that of the past. Fearing that the state would take the remaining pasture-land, the people began again to compete with the state. After discussions among the elders, they began to plough up the pastures, so that they would be recognized as the ongoing users. It was normally a taboo to hoe and cultivate pasture, but the only way they could find to save it was to cultivate it. A further threat to farmers was posed by Department of Agriculture workers, who made constant efforts to encourage people to unlearn old practices and learn new methods. Attempts to villagize the mountain settlements were worse still. Enset, the basic crop in mountain Gamo, needs up to five years to ripen and requires constant manuring. For this reason alone, moving house and leaving the enset behind were not feasible. The higher the elevation, the longer the time required to grow. The Gamo rejected the villagization scheme, but, when pressure mounted on some villages near Chencha, they set up secondary home-type bamboo structures but left them unthatched, explaining to officials that this would protect the structure from insects. The government collapsed before they had moved into the houses and the structures were abandoned. In the lowlands, villagization had already been introduced, in Lante and Ch'anno and Shelle, by the Ministry of Community Development in the late 1960s, under the label of 'community development'. Only Shara was villagized under the Derg. In 1978 the legal basis for the establishment of producers' cooperatives was set up. In 1985, 540 well-organized and productive cooperatives were incorporated into national planning, of which two were Ch'anno and Lante in the lowlands. There were nine other agri-producers' cooperatives, which were classified as being in the first phase of development during this time. At the same time the whole of Gamo was organized into 300 Peasant Associations.[19]

In the end, the military failed like all previous states. It failed for various reasons (external and internal), both in the implementation of its projects and in acknowledging and recognizing Gamo practices and forms of tenure. For the Gamo, the state was too powerful. It competed with its subjects for land, for priority in irrigation and so on. The only evidence people had of their ownership of land were receipts of tax payment. The major failure of the land reform was that it was not attuned to the various tenures that already existed in the country. Also it did not have an infrastructure capable of implementing its reforms. Although selling or buying land was forbidden by the land proclamation, in Gamo, Konso and some other areas rural land was sold and bought. Although the state was not able to implement the land reform totally, it did, however, succeed in establishing administrative structures to replace the earlier systems. Owing to the lack of institutional strength of the state, in most places in Gamo the new structure worked closely with 'traditional' institutions, not so much because this was desired by the state, but because they were too powerful to be neglected.

By 1990, most Gamo markets (there were more than fifteen big markets in the mountains alone) and over three-quarters of the settlements were run by local assemblies and led by elected elders who had no links with the state structures. These institutions were crucial for keeping law and order and for keeping life and property safe. When the Derg collapsed and the country was in near anarchy at the centre and in many other places, life in Gamo and in some southern communities continued undisturbed. When the new government that replaced the Derg proclaimed that it was a 'people's government' and when its cadres said that farmers would not be required to pay tax any more, the farmers had heard it all before and were not impressed; they can even be said to have reacted to the announcements with scepticism and suspicion.

Conclusion

Throughout the Gamo history of relations with state structures, land has been the battleground of power and for processes of identification. It has been a key means through which the state has attempted to interfere in and control Gamo lives. But the intense, meaningful and intimate link that Gamo people have to their land has

---

19. Ethiopian Mapping Authority (1988).

also been a source of strength. It has provided them with a source of resistance that has prevented them from becoming swallowed up by the patronizing rural policies that did not heed local knowledge or its significance for maintaining the landscape. Despite the power of the state, fully recognized by Gamo people, the Gamo have never relinquished their knowledge of how they should live and farm properly on their land. Nor have they forgotten the complementarity between highland pastures and valley homesteads, and its importance for sustaining their livelihoods. Overall, it has been their challenge to reclaim the lowland areas taken by previous governments and, by reinstating this link to their land, to regain a full sense of self, progressively 'sold' throughout the twentieth century.

Chapter 8

Identity, Encroachment and Ethnic Relations: the Gumuz and their Neighbours in North-Western Ethiopia

Wolde-Selassie Abbute

Introduction

The Gumuz[1] are a group of people that inhabit an area that extends from Metemma southwards through Gondar, Gojjam/Metekel, and across the River Abbay up to the Diddesa valley in Wollega, western Ethiopia (see Map I.1). They designate themselves the Bega.[2] Linguistically, they belong to the Koman group of the Central-Sudanic branch in the Nilo-Saharan language family (Blench 2000). In the present administrative context, the Gumuz predominantly inhabit Metekel Zone to the north and Kamashi Zone to the south of the River Abbay, in the Benishangul-Gumuz National Regional State. Metekel Zone is the focus of this study.

Low-lying topography dominates most of Metekel Zone. Climatically, the zone is classified into 82 per cent *qolla* (low-altitude), 10 per cent *woina-dega* (mid-altitude), and 8 per cent *dega* (high altitude) (MZAD 1999: 1). The 1994 census of Ethiopia estimates the population of the zone to be 201,521 and describes the number of major ethnic groups as 66,965 Gumuz, 47,900 Amhara, 32,037 Shinasha, 27,050 Oromo, 17,155 Agaw, 2,875 Kambaata, 2,179 Hadiyya, 725 Tigraway[3] and the rest composed of others (CSA 1996: 41). The zone is further divided into seven *woreda* – Dangur, Guba, Mandura, Dibate, Bulen, Wombera and Pawe Special Woreda. Except the last, which is inhabited by the state-sponsored resettler population of the 1980s and the recent highland immigrants of the 1990s,

---

1. In the written sources, the people are referred to by different ethnic and generic names such as *Bega, Gumis, Gumz,* Gumuz, *Say, Sese,* and *'Shanqilla'*. The local people call themselves *Gumz.* However, I use the ethnic term Gumuz to refer to the group as is it the most widely used term in the written sources and by the regional administration.
2. *Bega* is an emic designation of the Gumuz, used to refer to their group members. The two ethnic terms *Bega* and Gumuz are used interchangeably throughout the study.
3. The ethnic group name 'Tigraway' is also used in the statistical report of the 1994 Population and Housing Census of Ethiopia results for the Benishangul-Gumuz Regional State (CSA/PHCC 1996: 43).

the remaining six *woreda* are inhabited by relatively mixed ethnic groups of Gumuz, Amhara, Shinasha, Agaw, Oromo and Kumfel (Qollegna[4] Agaw). The autochthonous Gumuz, who are the most numerous ethnic group of the area, mainly inhabit the lowlands in all six *woreda*, whereas the others inhabit mostly the mid- and high altitudes. On the other hand, the state-sponsored resettlers in Pawe Special Woreda inhabit a lowland area of 1,000 masl to 1,200 masl, which was previously inhabited by an estimated 15,000 to 20,000 Gumuz (Agneta and Tomassoli 1992: 343), who were displaced by the resettlement scheme. According to the 1994 population and housing census, an estimated 35,858 state-sponsored resettlers inhabit the resettlement scheme area – ethnically differentiated into 26,900 Amhara (from Wollo, North Shoa, Gojjam and Gondar), 2,868 Kambaata, 2,178 Hadiyya, 1,286 Oromo, 1,110 Agaw (Wollo-Tigray/Seqota), 502 Tigraway and the rest others (CSA 1996: 43).

In order to facilitate the discussions in this study, the diverse ethnic composition profile in Metekel can be divided into four broad subcategories: (1) the autochthonous Gumuz; (2) the long-standing highland neighbours (Shinasha, Amhara (Gojjam/Gondar), Agaw/Gojjam and Oromo); (3) early immigrant settler peasants from Wollo (Amhara) since the 1950s; and (4) the state-sponsored resettlers of the 1980s. Since dealing with the entire web of inter-ethnic relationships among all categorical groups is beyond the scope of this chapter, I focus mainly on the ethnic relations of the Gumuz with their highland neighbours, emphasizing the effects of encroachment on the land resources of the former and their interactions with the latter. Methodologically, out of the seven *woreda* in the Metekel Zone, four were selected for field study. These are Dibati, Mandura, Dangur and Pawe Special Woreda, where severe encroachment pressure has been occurring on the resources of the Gumuz. Both site-specific fieldwork and exploratory assessments were made for this study, using a combination of data collection methods.

Premises for Understanding Identity and Ethnic Relations: Conceptual Considerations

Ethnicity, explains Ronald Cohen (1978: 397), is 'a series of nesting dichotomizations of inclusiveness and exclusiveness' where 'the process of assigning persons to groups is both subjective and objective, carried out by self and others, and depends on what diacritics are used to define membership'. Ethnic groups, he considers, 'are those widest scaled subjectively utilized modes of identification used in interactions among and between groups' (ibid.: 385). In addition, ethnic identity is considered to 'refer to a process by which individuals are assigned to one ethnic group or another', which 'implies boundaries, their creation, maintenance and change' (Kunstadter 1978, as quoted in Cohen 1978: 386).

Barth's (1969) ethnicity approach, writes Anthony P. Cohen (1994: 59), is an 'intellectually liberating paradigm shift' credited for taking 'ethnicity out of its social structural closet, and locating it firmly in the realms of the interactional, the

---

4. Qollegna means lowlander.

transactional and the symbolic'. According to Barth (1969: 15), the focus of investigation should be on 'the ethnic boundary that defines the group, not the cultural stuff that it encloses'. This statement highlights the central tenets of Barth's conceptual approach, where the main points of investigation are summarized in his preface to the 1998 reissue of *Ethnic Groups and Boundaries* as:

- That ethnicity is a matter of social organization above and beyond questions of empirical cultural differences: it is about 'the social organization of culture difference'
- That ethnic identity is a matter of self-ascription and ascription by others in interaction, not the analyst's construct on the basis of his or her construction of a group's 'culture'
- That the cultural features of great import are boundary-connected: the diacritica by which membership is signalled and the cultural standards that actors themselves use to evaluate and judge the actions of ethnic co-members, implying that they see themselves as 'playing the same game'. (Barth 1998: 6)

The ethnic identities of marginal groups, according to Fukui and Markakis (1994: 6), 'are essentially political products of specific situations, socially defined and historically determined'. Ethnicity, states Ronald Cohen (1978: 389), has no existence apart from inter-ethnic relations. Inter-ethnic interactions are patterned on an ethnic boundary dichotomization: in-group/out-group or 'we' and 'they' (Barth 1969). Ethnic identities are manipulated in situations of inter-ethnic relationships.

   The immediate generation of ethnological studies following his volume *Ethnic Groups and Boundaries*, states Barth (1994: 18), 'gave much attention to the linkage of ethnicity to the concept of niche, and to the theme of resource competition'. Emphasizing divergent adaptations, he explains, 'particular cultural traits may be useful as adaptations to particular environments and modes of subsistence' where 'resource competition between populations with distinguishing cultural features may provide a special impetus to their mobilization in collective action on the basis of shared ethnicity' (ibid.). Thus, the divergent ecological adaptations perspective, according to Barth (1969, 1994), provides a necessary component in the study of plural situations with competitive ethnic relations. In this study, the form and nature of ethnic identity and inter-ethnic relations are examined in the context of encroachment of resources of the autochthonous Gumuz by their highland neighbours.

Indicators of *Bega* and *Shuwa* Identity

In the local ethnic identity politics of 'inclusion versus exclusion' or 'we/us' and 'they/them', the complex identification process of 'Who is against whom and why?' and 'Who belongs to whom and why?' (using the expressions of Schlee 2000: 89) is explained here through an examination of ethnicity markers. The Gumuz make a strong boundary dichotomization between 'we' and 'they'. They refer to their group members as *Bega* (our people) and all other highland neighbouring members as

Shuwa[5] (other highlanders), which includes Amhara, Shinasha, Agaw, Oromo and all other highland settlers in Metekel. In the process of ethnic identification, several features are articulated to mark the sociocultural distinction of the Gumuz (*Bega*) identity against the highland neighbours (*Shuwa*). These include: the popular genealogical myth of common origin and historical life experiences; occupation of divergent ecological niches as lowlanders versus highlanders and the differing modes of livelihood; differing belief systems, marriage systems and patterns of authority; overt physical features such as pigmentation and bodily scarification; and other identification symbols such as language, costume and folklore. Moreover, various features of ethnic stereotyping are used to label identifications.

According to the popular origin myth of the *Bega*, once upon a time, twin sons were born from a mother. One of the sons was black and the other was fair. As they grew up, they competed for power in the presence of a local assembly. First, they were given a mule for racing. The *Shuwa* galloped well but the *Bega* fell from his mule. Then, they were offered to choose from a bow and arrow and a spear and shield. The *Shuwa* chose the spear and shield, while the *Bega* took the bow and arrow. Later, they were made to choose between porridge and *injera*[6] as their staple food. The *Bega* took porridge while the *Shuwa* took *injera*. Then, they were asked to settle either in the central highlands or towards the lowlands and the periphery. The *Shuwa* selected the highlands, to the right, and the *Bega* selected the lowlands, to the left. Thus, the Gumuz claim genealogically to have originated from a common ancestor with their *Shuwa* neighbours, but they differed in their choices right from the very beginning.

The historical aspects such as the ceaseless external encroachment, ruthless subjugation and enslavement of the Gumuz by the central highlanders are strongly emphasized in discussions that shape and maintain the *Bega* identity. The long experience of Gumuz of slave raiding by the highlanders forced the Gumuz to escape and retreat further into the inaccessible lowlands. Several Abyssinian kings conducted devastating and destructive campaigns to subjugate and enslave the Gumuz, who were frequently raided for their young men and women. According to the oral accounts of Mandura and Dangur Gumuz informants, the shocking experiences of the slave raids by highlanders, witnessed during the early years of some of the living informants, included the cutting of one of the enslaved's strong leg tendons to prevent escape. These horrible experiences of enslavement have a powerful influence on the consciousness of inclusion and ethnic boundary maintenance in the process of *Bega* identification. It constructs an inclusive *Bega* identity against the extreme exclusion of *Shuwa*.

Various ethnic stereotyping features are also used to label the *Bega* identity against *Shuwa*. The *Shuwa* refer to the Gumuz with a derogatory generic term, *Shanqilla*,[7] implying an inferior status. The Gumuz also in turn refer to the *Shuwa*

---

5. *Shuwa* is an etic categorical designation used by the Gumuz to refer to their highland neighbours. The terms *Shuwa* highlanders and 'highland resettlers' are used interchangeably throughout the study.
6. The round pancake-type staple food of central Ethiopia made from teff.
7. This derogatory term also implies darker pigmentation.

with the generic term of hatred, *Mittiha*, which connotes cruelty and brutality. Behavioural stereotypes are also often cited and used as an indicator of difference: the *Bega* consider the *Shuwa* to be cruel, heartless and inhuman people who used to raid and enslave them brutally. They are thought to be double-crossing people who have ceaselessly encroached their resources and infringed their rights. Behavioural practices such as telling lies and stealing were said to be totally unknown among the *Bega* before the intrusion of the *Shuwa*. The mistrust between the Gumuz and the *Shuwa* is noticeable in multiple contexts. For instance, according to the interpretation of my Gumuz research assistant, Wubalta Enno, during the fieldwork some Gumuz members in Mandura were sceptical of the effectiveness of the 1999 and 2000 polio vaccinations offered free of charge. Their suspicion was based on their perception that the *Shuwa* would not offer something of real value for free. Many Gumuz were sceptical of getting their children vaccinated, considering that 'had the medicine really had the stated preventive value, the *Shuwa* would not have given it free of charge'.

The different ecological niches used by the *Bega* and the *Shuwa* are also cited as one of the main attributes of their difference. Since the Gumuz inhabit the lowlands, their entire way of life is adapted to the hot climate, undulating gentle slopes, plains and valley bottoms. Shifting cultivation and the gathering of wild forest foods, hunting, fishing and honey collection form the main sources of subsistence for the Gumuz. The *Shuwa* mode of subsistence is based on mixed farming, plough cultivation and animal husbandry. Religion is also cited: the Gumuz are followers of a traditional belief system, whereas their *Shuwa* neighbours are predominantly followers of Orthodox (monophysite) Christianity, together with a significant number of followers of Islam. 'Sister exchange' marriage is the most dominant and standard form of marriage practised among the Gumuz, which is completely different from marriage practised among the *Shuwa*. Patterns of authority among the Gumuz are ideally egalitarian; patterns of authority among their *Shuwa* neighbours are considered to be hierarchical, based on either achieved or ascribed status.

Bodily scarification is one of the most powerful symbolic identity markers or emblems. For instance, removing the two central lower incisors by the Rendille of Kenya, Schlee (1994b: 132) explains, marks a highly visible identity emblem of the group. Similarly, among the Gumuz, bodily scarification marks make clearly visible identity symbols and emblems. The overt physical scars are also thought to serve as marks of beautification, especially in the case of women. Men mostly have scars on both the left and right sides of the upper cheeks; women mostly have decorative scars on their backs, cheeks, stomach and upper arms. The bodily scarification is performed on boys and girls at puberty and considered to be a mark of their ethnic identification. These bodily scars are referred to locally as *moqqota*.[8] According to

---

8. The traditional practitioner of *moqqota* is known as an *ittemoqqota*. The different bodily scars have different shapes and are referred to by different names. Bodily scars on cheeks are called *moqqotlissa*; scars on the back of women are called *moqqotbongua*; scars on the abdomen of women are called *moqqotaiila*; scars on the upper arm are called *moqqotqo'ea*; and scars on the lower arm are called *moqqota'ea*.

informants, a Gumuz without a proper facial scar is not considered *Bega*. Either (s)he is not *Bega* or (s)he is someone who was born outside the area from parents who were taken as slaves. Thus, these bodily scars articulate the *Bega* identity over and against that of the *Shuwa*.

Equally important overt cultural attributes, considered by the *Bega* as identity markers, are language, costume and folklore. Cultural attributes such as language, dress and ritual procedure are considered as merely symbolic labels denoting the different sectors of a single extensive structural system (Leach 1954: 17, cited in Schlee 1994b: 132). As stated earlier, linguistically, the Gumuz belong to the Koman group of the Central-Sudanic branch in the Nilo-Saharan language family, whereas their *Shuwa* neighbours belong to Semitic (Amhara), Cushitic (Oromo and Agaw) and Omotic (Shinasha) language families.

In their clothing, Gumuz women wear a cotton skirt wrapped around their waist that extends down to their knees and, most of the time, they leave the upper part of their bodies naked. Interviews with Gumuz in Mandura provide evidence that wearing the full dress of the neighbouring *Shuwa* women and the customarily worn shawl (*natala*) is perceived by the Gumuz women as deviant behaviour and as showing disrespect to their traditional way of life. Gumuz women wear strings tied tightly around their arms on two places below the shoulder and above the elbow. On special occasions women make leather strings cut from fresh skins of goats slaughtered ritually to celebrate ceremonial events and they tie them tightly around their legs. An example of such a ritually celebrated occasion is an event to mark a bride's virginity, in which the groom offers a goat to the singing, dancing and chanting womenfolk of his and the girl's natal village. Women also wear ornaments, such as strings of coloured beads around their necks, iron rings on their fingers, large earrings, fixed iron ornaments on their noses and bracelets on both their arms and legs. Girls at puberty wear over their skirts thin strings of leather tassels that are wrapped around their waists and suspended on their buttocks at the back as a decoration. These provide an attractive rhythm on the occasions of traditional dances and chants. They also smear ochre mixed with castor bean oil on their hair. The women carry goods balanced on a stick over their shoulders, in suspended nets tied at both ends, into which large gourds, baskets and bags containing the load are fitted. In contrast, the *Shuwa* women differ in their costume and carry loads on their backs.

On the other hand, the style of clothing worn by Gumuz men is not so easy to describe because of the influence of a combination of autochthonous, European, highland Ethiopian and Sudanese practices. However, adolescent youths wear local daggers inserted into a dagger sheath tightly tied together with packed strings of leather tassels suspended on either side of their waist. The dagger is used both for defence and for cutting meat, and the suspended strings of leather tassels make a percussive rhythm when singing and dancing.

Unlike most of their Orthodox Christian neighbours, whose priests are relatively familiar with written scripture, the Gumuz are primarily non-literate people who depend fully on oral traditions and are rich in folklore, such as poetry and prose narratives. Their folklore has been woven from the substance of their experiences and portrays their struggles with the mysteries of existence as well as their explanation about the unknown in order to create order and reason. Apart from its extensive artistic value, the folklore expresses their struggle for their rights and the wars and conflicts that they have had with their neighbouring ethnic groups. They also express their struggle to defend and safeguard their human and land resources, and their perceptions and explanations of the secrets of nature, life and death.

Intra-ethnic Relations

On the other hand, within the Gumuz, membership of a clan is the most important symbol of identification. The clan is composed of various neighbourhood members who claim to share patrilineal descent, and its territory is strongly defended by the members from outsiders. Inside a clan territory, members of closer kin groups live in villages.

Membership of a clan affects participation in death and mourning rituals (a male visiting clan member has to fire a bullet twice or more, based on his lineage closeness), in defending and avenging attacks on other clan members, and in all important life events such as birth, circumcision (men), marriage and other related ritual events. My Gumuz assistant Wubalta Enno is almost always asked about his clan background by most of our Gumuz informants, particularly in the first introductions. In other words, a Gumuz person without a clan membership does not exist. As earlier described, a number of clans form larger and broader territorial groups such as the Gumuz of Mandura, Dangur, Gublak, Guba, Dibati and Wombera. In turn, these different Gumuz groups form the much larger territorial category of the Gumuz of Metekel. In further extended order, wider categories, such as the Gumuz of Metekel, the Gumuz of Kamashi and the Gumuz of Metemma, form the entire Gumuz ethnic group as a whole.

Intra-ethnic clan interactions are crucial for the very survival and coexistence of the different groups as a community, and relations are held with mutual recognition according to their tradition. Violation of such recognition results in severe inter-clan feuds. Normally, the neighbouring clans in their respective territories[9] coexist peacefully, with a respectful recognition of each other in their interactions with individual members as well as among groups. At a lesser scale, intra-clan lineage relations are also handled with care. In their fairly egalitarian system, any Gumuz member is considered as fully equal to any other fellow member or neighbour. However, potential causes of intra-ethnic bloodshed include adultery, belief in the evil eye and insults. Adultery is considered to be among the worst of crimes and its punishments include the death penalty. When an individual belonging to another

---

9. Ideally, all clans occupy localized territories. However, in reality the neatly bounded clan territories are difficult to identify, and there are cases where members of the same clans are observed in different territorial units. According to Gumuz informants, clan members residing outside clan territories are those who joined other clans with the full consent of the latter, who accepted the former and allowed them to share their territories in a peaceful way. Even in such a case, the status of those who joined from outside is subordinate to the clan members who claim to be the 'true owners'. Hence, every Gumuz prefers to stay inside his clan territory unless otherwise forced to leave by circumstances such as intra-clan and inter-clan feuding.

clan commits adultery, the issue is by no means easily resolved. Showing overt contempt for an individual may also have serious repercussions. Abduction of girls has a serious risk of causing bloody hostility because getting a wife is possible only in exchange for a sister and a girl's virginity is highly valued. Moreover, a claim for the return of an exchanged sister or a replacement in the case of the divorce or death of a wife before bearing children causes inter-clan tension. An equally severely punishable practice is incest. I was informed by one of the local authorities of Metekel Zone of a case where several Gumuz men from Bulen Woreda were arrested in the zonal prison for their act of burying alive one of their male fellows. He had committed incest by making a young and close relative lose her virginity. They buried him alive, they said, instead of wasting a bullet shooting him. The practice of retaliation feuds that target any male member of each other's groups creates complex tensions that involve every male member of the conflicting clans and make the conflicts more protracted. Historically, intra-ethnic conflicts are resolved through the arbitration of community elders in a local institution called *mangima*.

Encroachment on Resources

The Gumuz view land resources as the sacred creation of *yamba* (the supreme deity) and consider it a symbol of their identity. Land resources are thus perceived to be sacred and cannot be alienated. Ideally, land cannot be sold, divided or owned privately, only its products. Land is believed to be the source of life as well as the final destination of life where people will be buried. Without land, it is thought that a clan or entire ethnic group will be lost. It is an ancestral heritage that has to be defended and preserved. Although the Gumuz have effective ways of indigenous resource tenure and management, they have also suffered encroachment on their traditional territories. The central state, under different regimes, has been reluctant to recognize their rights to their land. This has served as a root cause for hostility with their neighbours.

There are two types of land encroachments in Metekel over successive regimes – gradual and rapid. The gradual encroachment has been taking place for centuries, through the continuous push of the neighbouring highland plough cultivators and spontaneous settlers in search of cultivable land and pasture. The Gumuz are aware of this process, which has pushed them further into the lowland peripheries. Both the local oral historical traditions and written accounts reveal that the Gumuz used to live in parts of the highlands of the present Gojjam (see later).

During the imperial regime of pre-1974, the Gumuz in Metekel were subjected to the gradual encroachment of the highlanders and spontaneous settlers, who expanded at an increasing rate beyond the Kar mountain range, which had previously served as a frontier between the Gumuz and the highlanders in areas neighbouring the Beles Valley. Alongside this process, new forms of migration began in Metekel from the 1950s. Higher numbers of immigrants than ever before trekked into Metekel from Wollo, as a consequence of environmental degradation in their home areas. According to informants, the voluntary immigrants from Wollo made agreements with the local Agaw feudal chiefs at Chagni (who were also charged with the affairs of the Gumuz), and obtained access to fields for cultivation and settlement. This agreement included an annual tribute payment to the Agaw chiefs. They were allocated land around Chagni, Dibati and Bulen, displacing the native Gumuz. This resulted in bloody conflicts between the Muslim Wollo settlers and the Gumuz. Due to their intensive ox-plough farming system, which demands smooth field plots, as opposed to the shifting cultivation system of the Gumuz, the Wollo settlers placed tremendous pressure on the local environment, and the Gumuz gradually moved down to the less attractive lands.

The gradual encroachment upon the land resources of the Gumuz by the neighbouring highlanders and spontaneous settlers continued throughout the Derg regime (1974–91). In addition, two forms of rapid encroachment occurred during this period. The first was the the Beles state farm, established in 1978/79, and the second was the 1980s government-sponsored emergency resettlement scheme. The Beles state farm was established in 1978/79 as part of the expansion scheme of state farms under the Birr-Humera Agricultural Development programme. The farm is located in the Dangur Woreda of the zone. Initially, the farm consisted of 500 hectares of land; in 1986/87, this was extended to 1,161 hectares; and there was an unfulfilled plan to add 1,000 hectares in 1987/88.[10] The farm ceased operations after the 1991 change of government.

The state-sponsored Beles Valley conventional resettlement scheme was set up in Metekel to overcome the emergency needs of the population affected by the 1984/85 drought and famine. It is one of the biggest resettlement schemes implemented in the country. The scheme is situated in the basin of the River Beles and covers an area of around 220,000 hectares. It hosted a resettler population of famine victims from northern Ethiopia and those who suffered land shortages in south-western Ethiopia. At its peak in 1987/88, the population involved in the programme reached a total of 82,106 (the household heads among them numbered 21,994; the family members 60,112). At present, the former resettlement villages and the newly emerging small towns in the area have merged, forming twenty *kebele* administered under Pawe Special Woreda. The ethnic composition of the resettled population is heterogeneous, as already stated earlier. In their original areas, north-central resettlers were mainly intensive cereal cultivators, while south-western resettlers were mainly enset (*Ensete ventricosom*) and root crop cultivators. In the resettlement scheme areas, these different ethnic groups with diverse backgrounds represent a microcosm of the cultures of the entire country.

The Ethiopian authorities considered the Beles Valley area in Metekel to be previously 'unoccupied/uninhabited virgin' lands with agricultural potential. However, before the arrival of the resettlers, the present Beles Valley resettlement area was inhabited by an estimated 18,312 Gumuz (Metekel Zonal Agricultural Department (MZAD) 1992). As already noted, Agneta and Tomassoli (1992: 343) estimate the figures of Gumuz to be in the range of 15,000 to 20,000. As a result of the implementation of the scheme, the Gumuz were pushed further to the peripheries of inaccessible forests and land, depriving them not only of their settlements and fields but also of their hunting and fishing areas. The infrastructural

---

10. MZAD (1992: 11).

aid that was later extended to rehabilitate the resettlers in view of developing the economy of the area completely neglected the Gumuz. Nothing was done to compensate them for their losses or to integrate them into the social development processes. Moreover, the influx of people coming to the area looking for different opportunities has enormously increased the pressure on resources, leading to deforestation and the destruction of the flora and fauna. The Gumuz livelihoods have been depleted.

Both the gradual and the rapid encroachments continue during the present regime (1991–). At the time of writing, a considerable amount of land is being leased out to private investors. The land of the Gumuz in Metekel is granted in lease rights to investors at Assosa, the regional state capital, which is located very far from the zonal capital, Gilgel Beles. Besides, the Ethiopian Investment Incentives Regulations No. 7/1996 granted incentives for investors (in this relatively underdeveloped area) to encourage them to engage in other pioneer investment activities. These include an exemption from custom duty tax on imported machinery and accessories, as well as exemption from income tax for five years. These pioneer investment activities include irrigated agriculture, agro-industry and manufacturing. Moreover, the same regulation grants similar exemptions for investors engaged in promoted activities including rain-fed agriculture, livestock farming, hotels and tourism, transport and storage services. The implementation of these regulations has been without adequate precaution or monitoring of its effects on the human and natural environment. In Metekel zone, seven investors who engaged in agricultural activities have been given 65,000 hectares of land, which was originally covered by forests of dense woodland mixed with natural bamboo (MZAD 1996, cited in ECA 1998: 27). Of these investors, four are in Dangur (40,000 hectares), one in Pawe (5,000 hectares), one in Mandura (10,000 hectares) and one in Guba (10,000 hectares) *woreda*. The allocation of land and forest resources to investors without thorough topographic, soil, vegetation type and other socio-economic investigation has had serious socio-economic and environmental consequences.

The encroachment has increased population density and has decreased the size of the land owned by the Gumuz community. The decrease in the landholding size has also caused a decrease in field fallow period. Since a short fallow field cannot rejuvenate well, the crop yields have decreased. Moreover, the influx of encroaching migrants has increased the livestock population. The increased livestock grazing and browsing inside the remaining forests have caused soil compaction and destruction of bamboo shoots, which serve as a food source for the Gumuz. The ever-increasing pressure on the land resources has intensified forest destruction. Above all, the Gumuz informants complained seriously about the increasing enmity with the migrants and ethnic conflicts, frequently costing numerous human lives on both sides. The Gumuz feel that their voice has rarely been heard and that their resource rights have not been considered or recognized. This has led the Gumuz to retaliate against the encroachers, while at the same time retreating further down into the lowlands. According to the Gumuz informants, it is usually after a heavy attack against the enemy and mostly at night that they retreat further into the inaccessible forests in fear of more organized and better equipped raiding from the highlanders.

Inter-ethnic Relations: the Gumuz and their Neighbours

Whenever explaining the uneasy coexistence with their *Shuwa* neighbours, there is always an emphasis on mistrust of the *Shuwa* on the part of the Gumuz, owing to the factors discussed above. Although the Gumuz use the broad category of *Shuwa* to denote those 'highlanders' they consider to be different from themselves, there is also a difference between the relations of the different groups within the *Shuwa* category and the Gumuz. Below, the relations of Gumuz to each ethnic group are delineated briefly, with special emphasis on the relationship between the Gumuz and the state-sponsored resettlers.

Gumuz and Amhara

The relations of Gumuz and Amhara are characterized by hostility and seems to be more polarized than that with any other group. Gumuz informants emphasized their historical experience of slavery and the repeated invasion and plunder by the Amhara armies under different regional monarchs and the central state during the imperial regime. After the sixteenth century, the centre of state was in Gondar, and the Gumuz of Metekel were very close to its influence. In order to bring the land, its resources and the people under control, Taddese Tamrat (1988: 12–14) describes how kings such as Sarsa-Dengel (1563–97), Susenyos (1607–32), Fasiledes (1632–67), Yohanes (1667–83) and Iyasu the Great (1683–1706) successively conducted devastating and destructive campaigns against the Gumuz. Finally they were able to incorporate them into a system of indirect rule by appointing neighbouring Agaw chiefs over them. This lasted up to the overthrow of the imperial regime by the Derg in 1974. Wendy James (1986) considers the accounts of early scholars such as James Bruce, Henry Salt and Charles T. Beke, who witnessed the fact that the Gumuz lived in the highlands of the present central and southern parts of Gojjam in the eighteenth and early nineteenth centuries. These accounts try to point out that, in the process of the exploitative predatory expansion of the highland Amhara, the Gumuz were forced to move gradually down to the lowlands in the far western periphery of the country in search of greater safety.

Gumuz and Agaw

The relation between the Gumuz and the Agaw is likewise characterized by conflict. The Gumuz were ruled indirectly under the Agaw chiefs appointed by the Amhara. According to the Gumuz informants, the Agaw raided them for slaves, pushed them away to the peripheries, exploited them and plundered their property (in the name of land tax), directly encroaching upon their land. In the nineteenth century, Charles Beke recorded an Agaw tradition that asserted that the Gumuz were the previous occupiers of Agaumidir, but had been displaced by the Agaw (Beke, quoted in Pankhurst 1997: 91). The Gumuz that I spoke to said that, as close neighbours, the Agaw have been able to approach the Gumuz cleverly: they have learnt to speak the Gumuz language and to maximize their material gains from the relationships subtly. Because of this, the Gumuz perceive the Agaw as 'deceivers'.

Gumuz and Qollegna Agaw

On the other hand, there is another group neighbouring the Gumuz locally known with a categorical name *Kumfel* (a term considered to signal contempt and stigma). The name preferred in the emic perception of the group is Qollegna Agaw. They are well known locally for their traditional honey production, which has a long history, and trees with several beehives are known under family names and are passed down through generations. Since they rely on the forests for their honey production, their activities are not harmful to the forest resources. The relations of Gumuz with *Kumfel* need further investigation, but no case of severe conflict between the two groups has been noted. Perhaps the lack of conflict is related to the way in which the *Kumfel* are ascribed a low status similar to that of the Gumuz. The name *Kumfel* is a pejorative categorical term ascribed by the dominating ruling majority in the past, although it is not as discriminatory a term as *Shanqilla*.

Gumuz and Shinasha

The relations between the Gumuz and Shinasha are said to be fraught with tension and they have an uncomfortable coexistence. Both the Shinasha and the Gumuz informants agree that they had severe conflicts in the past and that these drove the Gumuz from locations that are now inhabited by the Shinasha. The Gumuz attribute the lower intensity and frequency of hostilities with the Shinasha to the latter's mainly hoe-cultivation system, which has less adverse effects on the forest resources than plough cultivation. However, the Gumuz still view the Shinasha with suspicion, and point out the latter's similarity to and close association with the Amhara, Agaw and other 'highlanders'. The Gumuz perceive the Shinasha's present alliance with them in local politics as a strategy through which they hope to gain access to Gumuz resources.

Gumuz and Oromo

The relations between the Gumuz and Oromo (who inhabit three of the seven *woreda* in Metekel Zone, Dibati, Bulen and Wombera) are almost similar to that of Shinasha, but with reportedly less tension and hostility than with the Agaw and Shinasha. Although the Gumuz do not attribute equally brutal treatment to the Oromo (like that of Amhara and Agaw), they are again sceptical of them. The basic factor contributing to their uneasy coexistence is again their encroachment on land resources and the difference in their relative positions of strength and weakness in the prevailing power politics in the area.

Gumuz and Wollo-Amhara Spontaneous Settlers

The relations between the Gumuz and spontaneous settlers have been the most hostile, resulting in frequent bloody conflicts since the latter's emigration from Wollo to the area started in the 1950s. The ceaseless immigration of the Wollo peasants (especially in areas inside Mandura, Dibati and Bulen Woreda of Metekel) pushed the Gumuz away from their land resources. In addition, the different farming system of Wollo peasants contributed to their worsening relations. Since the Wollo peasants usually feel insecure because of their numerical minority, they regularly invite more immigrant countrymen both for security and economic survival, which in turn exerts more pressure on the land resources of the Gumuz. In this inimical relationship, the two sides pursue retaliatory feuds that exacerbate the rift between the two groups. Regular ethnic conflicts between the two have been common since the Wollo settlers' early immigration days. However, the worst of all conflicts is the one that occurred in the years from 1992 to 1994 following the political change of the country in 1991. This ethnic conflict claimed numerous lives from both sides and destroyed enormous amounts of property, displacing around 10,490 Wollo immigrant settlers (Wolde-Selassie 1996: 11).

The conflict was mainly attributed to vindictive measures taken by the Gumuz for their past losses, influenced by the new state policy on ethnicity, which structured administrative boundaries along ethnic lines. Its adverse impacts on human lives and property has been described by Berihun, who based his accounts on the reports of immigrants:[11]

> According to the estimation of some of my Wollo informants, about 329 persons were killed and elders, children, and disabled persons were burnt alive with their houses. A total of 6,833 rural houses, 185 mosques that are made of grass roofs, one church, five elementary schools and a service cooperative shop were burned during the conflict. About 1,792 cattle and too many sheep and goats were looted. In addition, the harvest both stored at home and at the field was destroyed (Berihun 1996: 119).

Many of these evicted farmers have migrated to the Beles Valley resettlement area, seeking better security near the Metekel Zonal Administration Centre, as well as looking for adequate farmland and pasture for their livestock. Most of the displaced immigrant farmers were settled in areas abandoned by state-sponsored resettlers of the 1980s, while many others also settled on the peripheries of existing settlements, clearing forests and making various institutional arrangements with the resettlers found there.

According to informants, especially in Dibati Woreda (some of whom were part of the following peacemaking process), this ethnic conflict was mediated and resolved by the involvement of a selected multi-ethnic council of elders, facilitated and monitored by Ethiopian People's Revolutionary Democratic Front (EPRDF) cadres. Under this broader elder's council, supportive subcommittees of elders were established in the respective peasant associations of the entire ethnic conflict area.

Gumuz and state-sponsored resettlers of the 1980s

The relations between the Gumuz and state-sponsored resettlers are characterized by hostility and violent conflicts that have claimed lives on both sides. The situation has

---

11. It should be noted that the casualty and property damage figures described refer only to those reported by the immigrants. Similar accounts estimating the effects of the same conflict on the Gumuz are not available.

deteriorated as more and more land has been claimed by the scheme. The Gumuz began expressing their grievances by attacking the resettlers located in the villages bordering their area. Low-level conflict continued throughout the Derg period, when armed resettler militia guarded the entire frontiers of the scheme area bordering the Gumuz.

Apart from the frequent casualties that occurred since the resettlers' first arrival, the volatile relations between the two groups broke out into open ethnic conflict that claimed a considerable number of human lives on both sides immediately after the 1991 political change in the country. According to information from different informants, the first conflict, which was carried out on an unprecedented scale, occurred on 27 December 1991 (18 *Tahsas* 1984 EC) in L4 village, where the Gumuz killed many Hadiyya and Kambaata resettlers. The intention was explained to be the expulsion of resettlers from the lost lands of the Gumuz. The next night, on 28 December 1991, the Gumuz again attacked Kambaata and Hadiyya resettlers in village L3 while they were at prayer in a Protestant church. Again this resulted in numerous casualties and many wounded.

The next heavy conflict occurred on 11 September 1993 (1 *Meskerem* 1986 EC), in which Gumuz gunmen attacked the L4 village market, resulting in several deaths and injuring many others. The actions were attributed to acts of local Gumuz and Shinasha leaders of the then zonal administration. On this pretext, the resettlers organized themselves and took retaliatory action on 22 September 1993, which resulted in a loss of more lives among the Gumuz. The retaliatory action first targeted the Gumuz and Shinasha administrative officials and then continued by plundering the Gumuz villages in an estimated eight peasant associations bordering the resettlement scheme area.

The situation was then brought under control. Since then, no conflict of a similar magnitude has occurred. But, as a consequence of this ethnic conflict, there has been a massive evacuation of the resettler population from the area. The number of resettlers, which in 1987/88 was 82,106, had gone down to 26,660 in 1993/94 (Wolde-Selassie 2000: 422).

This severe ethnic conflict was resolved, at least for the moment, through an extended peacemaking process. It succeeded especially because of the relentless efforts of selected elders from both parties, which were also facilitated by the EPRDF cadres. The resettlers designated the peacemaking process as *shimgilinna*. However, the term *mangima* is maintained because the accompanying ritual performance was conducted, applying the customary peacemaking practices of the Gumuz in order to communicate better with the latter.

Inter-ethnic Integration

Despite the dominance of conflictual relations and uneasy coexistences, there are also some existing and emerging symbiotic relations across ethnic boundaries. The contexts of cooperative inter-ethnic relations include local market exchanges, bond friendships, sharecropping arrangements, religious conversions and neighbourhood and other similar local informal institutional arrangements.

Exchange of products in the local markets is an important component of the Gumuz livelihood. Most of the Gumuz bring their own produce to local markets and sell them to direct consumers or traders. These in turn take the goods out of the area to an extended network of sellers. While there is exchanging for basic subsistence, no trading for accumulation has been observed among the Gumuz. The local market exchange places are existing, emerging and expanding centres of symbiotic inter-ethnic interactional contexts between the Gumuz and their neighbours. The Gumuz informants in Mandura and Dangur Woreda stressed that the improved local market exchange had brought about a most important positive effect. Although most of the services are not primarily targeted to their benefit, the Gumuz also noted considerable improvements in infrastructure and services such as transportation, potable water, education, health services and grain mills. These changes have served as a context for integration and mutual interaction, directly or indirectly.

Mutual bond friendship between the Gumuz and their neighbours is also an important existing and emerging form of symbiotic interaction across ethnic boundaries. These individually established acquaintances have been further extended in a network of friend-of-friends relationships between neighbouring ethnic groups. For instance, *Bega* and *Shuwa* friends intentionally visit each other. They enjoy each other's company during important life events such as happy and sad occasions like festivities and mourning. Those in need of loans of cash and grain get easy access through their friendship relations. Through friendship, the *Shuwa* get access to sharecropping arrangements with the Gumuz. Bonded friends act as nodal figures in channelling information and resolving misunderstanding and conflicts. However, the Gumuz seem to be sceptical of the trustworthiness of their friendships with their neighbouring *Shuwa* and also consider these 'friendships' to be the basis of unequal exchange mostly aimed at exploiting the Gumuz.

The traditional *michu* institution is considered to have come to Metekel with the Oromo, and is the most intimately binding friendship institution ritually established between members of the same or different ethnic groups. Mostly, *michu* is established between people without close blood relations. Members who want to establish *michu* will first assess each other's qualities. Then they organize a feast and state their wish for respected elders to process the ritual, calling for the company of their other close members. On a fixed date, the right-hand index fingers of the two friends will be pricked. Then, the two hold tightly each other's fingers so that the blood of one will enter into the other. After this, the two friends are considered to have the same blood. The different ethnic groups of Gumuz, Shinasha, Oromo, Agaw and Amhara have established *michu* between their interested members, especially in Dibati, Bulen and Wombera Woreda of Metekel. Once *michu* is established, the two families are considered relatives and cannot intermarry. They share close relations, one seeking the company of the other in most important life events. Above all, the two *michu* families play key roles in resolving conflicts and mediating in misunderstandings that emerge at different intra-ethnic and inter-ethnic group levels.

Sharecropping arrangements between the Gumuz and their neighbours in general and highland migrant resettlers in particular are practices that have existed at different levels in the past and in recent years. In areas away from encroachment pressure, it is least practised or not known at all, while it is widespread in Gumuz areas with high encroachment. Here, the limitation in land has introduced among the shifting cultivators claims to ownership of an earlier cleared field left under fallow. Therefore, mostly it is this fallowed field which is given by the Gumuz claimant to a highland migrant ox-plough cultivator. The usual terms of a sharecropping arrangement are an estimated one hectare of farmland rented in exchange for an average 200–300 kg of grain (mainly finger-millet or sorghum in the case of Beles Valley), depending on the potential of the land under cultivation. The Gumuz and the migrant plough cultivator first agree the details of the terms of exchange. Then the final approval and decision are made in the presence of Gumuz elders. After harvest, the sharecropping partners share the products on the basis of the terms fixed in their earlier agreement. They extend their relations further and share the company of one another, establishing important networks of relationships between their communities.

The Gumuz have been influenced by the religions of their neighbours to a differing extent. For instance, the Gumuz neighbouring the dominantly Orthodox Christian highlanders have been influenced by their religion. Islam has also influenced the Gumuz near the Sudan border. On the other hand, the Gumuz of Metekel bordering the present Kamashi zone of the previous Wollega region along the River Abbay are influenced by elements of evangelical religion from the south. The Gumuz converts of Orthodox Christianity seem to oscillate between the practices of their traditional belief system and the demands of the new religion, and are not considered as strong adherents of the latter. Usually, the *Shuwa* neighbours establish local religious institutional relations in the form of godparenthood or fictive parenthood and as members of religious associations, such as *mahber* and *senbete*. In most cases, through such institutional arrangements, the *Shuwa* neighbours get easy access to the land resources of the Gumuz, which the latter are still sceptical about even in such contexts as the Gumuz do not feel that they have been considered as equals.

Intermarriage between the Gumuz and *Shuwa* neighbours is not yet common. The only inter-ethnic marriage observed between the two groups to a noticeable degree is one that has been establishing between a Gumuz man (usually an elite townsman) and Agaw woman. Gumuz women rarely marry with neighbouring *Shuwa* men. Those Gumuz men who have married Agaw women gave one of the reasons for the marriage to be that they did not have a sister to give in exchange for a Gumuz woman. Intermarriage with Agaw women signals positive interactions between the two groups, who have lived for a longer period neighbouring each other.

The Role of the State

Although dealing deeply with the role of the state in the complex issues of inter-ethnic relations is beyond the scope of this chapter, some points with direct relevance here are worth considering. The present government seems to have instrumentalized ethnicity in the country in general and in Metekel in particular. Metekel, as explained above, is a context of complex ethnic relations. However, the state's ethnic policy does not seem to have a clear knowledge of the extent of its complexity at either the conceptual or the practical level. In such a multi-ethnic context, local physical administrative boundaries are demarcated along ethnic lines without serious consideration of the fluidity of ethnicity on the ground. Hence, the state's ethnic policy can be considered partly responsible for the bloody inter-ethnic conflicts. This is seen particularly in the case of Metekel, where a major political alliance has formed between Gumuz and Shinasha. These are both considered as the major local 'native' ethnic groups, although within this there are still complex politics. Primarily the two ethnic groups share major local administrative positions. This has created implicit and explicit resentment by all other resident groups of the area. There is a deep-rooted wider rift between those who are considered as the 'autochthonous' inhabitants of the area and those considered as newcomers from outside (*matte*).

However, after suffering the worst effects of the conflict, the same state authorities have attempted to resolve them. The role of the state authorities was stressed by informants, particularly in their facilitation of local elders' efforts at peacemaking through their traditional institutions. Thus, as a result of the tireless efforts of traditional elders, the ethnic conflicts have been somewhat resolved – at least temporarily. Mainly, it is after this important landmark role of elders in ethnic conflict resolution that the state authorities seem to have realized the power of a local solution to such sensitive ethnic problems. The efforts of the state authorities, especially in their provision of the necessary logistics, should not be underestimated. In certain cases, my Dibati informants reported that the local political authorities have provided animals to be used in rituals. For instance, the oxen slaughtered at the Chagni peacemaking ritual (to resolve the conflict between the Gumuz and Wollo immigrants) were reported as being provided by the local branch of the ruling EPRDF party. However, in such an ethnically complex setting, the role of the formal state institutions in the issues of ethnicity and ethnic relations is far outweighed by the role of the customary informal institutions. In matters pertaining to ethnic relations, the local authorities in all my study *woreda* informally confirmed the fact that, had it not been for the relentless efforts and roles of community elders through informal institutional arrangements in differing settings, they would not have managed to handle issues related to such a complex and fluid subject as ethnicity at all. In other words, they implied that the role of the state in ethnic relations without the primary role of customary institutional arrangements is futile.

Conclusion

The excessive encroachment of the highlanders and the resultant competition over resources have served as one of the root causes for the prevalent inter-ethnic conflicts and uneasy coexistence between the *Bega* and their *Shuwa* neighbours. On the one hand, the Gumuz predominantly signal deep-rooted mistrust and enmity towards the *Shuwa*, owing to the latter's reportedly contemptuous, expansionist and exploitative relations across political regimes. On the other hand, despite the predominant conflicting relations, there is also some existing and emerging integration across ethnic boundaries that may provide the basis for more positive relations in the future. These more positive relations are based on customary and new forms of inter-ethnic communication and alliance, but are also supported by state institutions. In this area the relations between different groups are tense, based as they are on long histories of inequality and competition for resources. This chapter has described a partnership between the state and local institutions that are able to mediate between different groups. Such partnerships need to be encouraged and developed further, in order that the bloody conflict that has characterized this area in the past may not be so common in the future.

Chapter 9

Debates over Culture in Konso since Decentralization (1991)

Elizabeth E. Watson

Introduction

This chapter examines some of the changes that have taken place in Konso since it became a 'special *wereda*', a self-administering unit in the new Ethiopian federal state that is structured along ethnic lines. Konso qualified for this status because it was considered to be a culturally and linguistically distinct minority nationality. In recognizing the Konso as a separate identity and giving them the right to a degree of self-determination, the government was also valuing their culture and livelihoods, reversing decades of government policy that celebrated northern Amhara culture and viewed all others as inferior. The rights of different nationalities are now enshrined in the constitution as: 'Every Nation, Nationality and People in Ethiopia has the right to speak, to write and to develop its own language; to express, to develop and to promote its culture and to preserve its history' (Federal Democratic Republic of Ethiopia 1995: 13; Article 39: 2).

Here, I examine the outcome of this process of decentralization. I explore how and why Konso was able to present itself as being an individual nation, and the impact that such a process has had on local understandings of themselves and their culture. Following an introduction that sets out the context and outlines some theoretical issues, I examine the changes that have taken place since 1991. First I examine this from the perspective of elites who have been involved in the new administration of Konso. Secondly, I examine the perspectives and experiences of the people at village level, who are supposed both to benefit from and implement the changes on the ground.

Konso and their Neighbours

The Konso people number approximately 215,000 and live on and around the Konso highlands in south-west Ethiopia. Ideas about the distinctiveness of Konso as a people, emic and etic, have focused on their hard work, and more specifically on the way in which they work together to produce the neat dry-stone walled terraces that cover most of the Konso highlands (Amborn 1989; Watson 1998). The Konso take great pride in their landscape and the hard work that goes into producing it. The flagpole at the centre of the roundabout at the main road junction in Konso has a sign on it – placed since the 1991 change of government – proclaiming in Amharic that 'the hardworking nature of Konso is known to the whole world'. The recognition of the challenges of work is exemplified in blessings that are given by older people to encourage the young: *coconado!* (have courage!) or *jabado!* (be strong!) When I went to Konso to start researching the agriculture of the area, I expected to have to justify my choice of region to the people. But, when I told one man what I wanted to do, he replied, 'Of course [you have come to the right place] – we are the best!'

The differences of Konso from their neighbouring peoples, the Boran, Guji, Burji, Dirashe (Gidole), Gawwada and Tsemai, are expressed and seen in the Konso food preferences, in the dress of men and women, in the funerary statues that are made for great men (and sometimes women) and in the architecture of the densely populated Konso villages. The villages of Karate region of Konso are usually located on easily defensible places such as the spur of a hill, and in the past were often walled to keep out wild animals and invaders. The distinctiveness is also thought by Konso people to be self-evident in the clan system, the generation-grade system, the way in which each village and area is administered by the 'father of the drum' and the division of all people into either farmers (*etanta*) or craftsmen/traders (*xawd'a*) (Hallpike 1972; Amborn, this volume). All men and women define themselves in relation to these institutions, structures and practices, which should not be assumed to be unchanging or constant throughout Konso. As well as change from internal dynamics, many aspects of Konso culture have been influenced by political and religious changes, particularly the impact of Protestant Christianity (Watson 1998). Despite this, 'being Konso' is part of a person's identity that is generally taken for granted and thought unproblematic. Various aspects of the elements described above are cited to justify this, and most of the aspects of cultural practice above are also frequently explained because they are 'the Konso way of doing things'.

There are also strong senses of affinity with their neighbouring peoples. The commonalities that are emphasized include ideas of shared origins and intermarriage (see Kellner, this volume, for discussion of relationship with Burji) and clan affiliations (for example, with the Boran and the Gawwada). Evidence that the languages are similar is also used anecdotally to stress the closeness of the relationship between the groups. There is a division of labour between many of the pastoralist groups and the agricultural Konso, which means that each can supplement the others' livelihood. In the past, Boran and Tsemai provided milk and meat-based products and livestock for Konso. In return the Konso farmers provided grain, and the Konso craftspeople provided iron and other objects, including the ritual iron objects used in Boran ceremonies. Such forms of cooperation also existed with other groups.

The way in which the relations between the different groups were characterized by cooperation and conflict and ideas of similarity and difference can be seen in the wooden funerary statues of Konso, *waka*, that mark the graves of great men (often clan leaders). In the funerary statues, the great man is depicted with the wives he has married, and also the people whom he, or his descendants, have killed during his lifetime (see Plate 9.1). The people he has killed, often Guji but also soldiers (who can be identified by their magazine belts), and those wives who are non-Konso can be identified by their hairstyles and dress, carved into the wooden statues. Conflict today is not common, but it is intermittent, especially in the border areas of Konso.

The idea of difference from and similarity with others sits easily with most Konso people. The same logic of thinking is applied within Konso: although there is

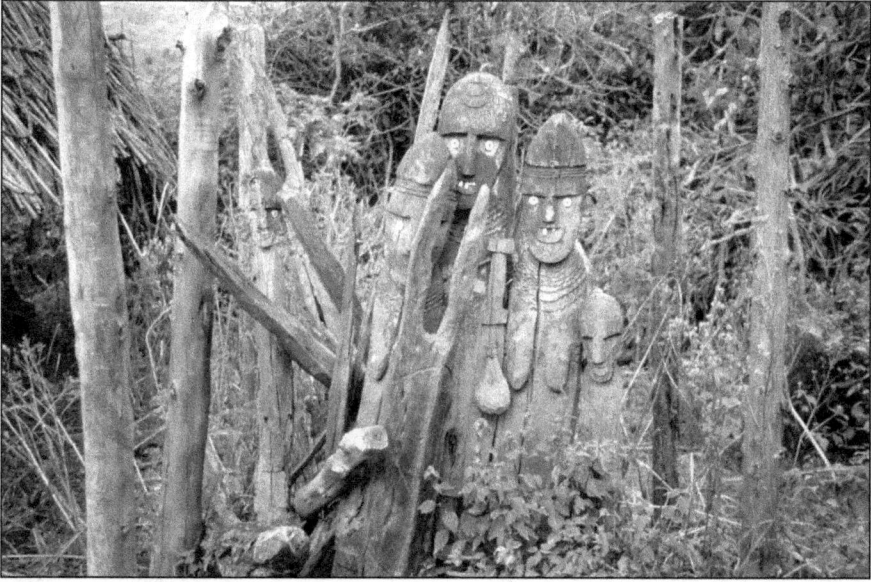

Plate 9.1 Konso *waka* (photo: E. Watson)

a wide area of Konso people who think of themselves as sharing a culture and a language, there are many variations in that culture and language between regions and villages. The differences between the regions of Turo, Karate and Kenna (sometimes known as Takadi) include variations in the number of generation grades in a generation-grade cycle, and a variation in the number of years that a person stays in a particular generation grade (Hallpike 1972). There are variations in dialect between these regions, to the extent that different words are used for the days of the week and the months of the year. Between villages, the roles and responsibilities of the institution of *apa timba* (the 'father of the drum') vary: generally an institution for maintaining order and fining offenders, his age classification and degree of ritual and administrative responsibility vary.

As with other identities, what is understood locally as being Konso is thus a multidimensional and multi-scalar matter. Konso people seem to have little difficulty in moving from understanding who they are on different scales and in different contexts. The elements of 'being Konso' that I have outlined briefly above[1] can be invoked in different contexts to explain different situations, to strengthen relations or to undermine them. The coexistence of different practices, ideas, histories and rituals is something that does not cause difficulty in most cases. Any contradictions that exist between them come to light in discussions, and in discussions any contradictions can usually be explained, resolved or overlooked. Had they been documented this would be harder to achieve.

---

1. For more detail see Hallpike (1972) and Watson (1998, 2002).

The ideas of similarity and difference, which come out strongly here, are central to ideas of self and other. Theorists have discussed the discursive construction of the 'other' as different as being crucial to the development of ideas of the self (see, for example, Hall 1996, for review). Barth's discussion of the construction of ethnic groups and identities also pointed to borders as the places where the work of identity construction takes place, as this is where the different identities and practices of other groups are encountered (Barth 1969).[2] This case of Konso points to the importance of difference, but shows that what is understood as 'being Konso' is only one element of identity that exists. There are many encounters with difference at many different scales, from personal to group relations. This leads to multiple identities, many of which cut across each other: people can be simultaneously of a particular clan, area, neighbourhood, ethnic group or gender. This allows different relationships to be mapped out in different contexts. The different forms of identities are therefore negotiable and also fluid. The fluidity of identity undoubtedly has advantages, as it allows alliances as well as conflicts to be formed across groups. The questions to be examined here are: have the drawing of an ethnic boundary around Konso and the prioritization of Konso culture and ethnicity as a definer of identity changed the way local 'culture' is understood and identity is expressed and experienced? Have they undermined the fluidity and how have they changed the nature of that identity?

The Wider Ethiopian Context

After the Derg was ousted from its control of the country in May 1991, the state was reorganized into several semi-autonomous regions defined by ethnicity (Young 1996). Under the previous regimes of Haile Selassie and the Derg, ethnicity was played down in an attempt to unite all peoples as one in Ethiopia. Both of these periods were dominated by a strong central state and by the Amhara culture, 'whose language has been given official status, and their religion and culture afforded an elite status' (Krylow 1994: 231). In response to this political and cultural domination, resistance to the strong state took the form of local nationalist movements, which were based on ethnicity (Pausewang 1994; Joireman 1997). When, in 1991, the Ethiopian People's Revolutionary Democratic Front (EPRDF) came to power, 'the role of ethnicity could not be ignored' (Joireman 1997: 397). Instead of subsuming ethnicity into a wider pan-Ethiopian nationalist project, the new state made ethnicity its structuring principle.

The dominant reason for centralizing ethnicity – or 'nationality'[3] as it is referred to – in the structure of the state may have been because of the important role that

2. By looking also at ideas of similarities I would also like to suggest that the form of identity and experience of individuals and groups is also important. In other words, a focus only on the 'other' and on borders, shifts the focus of identity construction outwards; I believe that there is also an internal dynamic to identity construction that should not be ignored.

3. An ethnic group could be considered a nation if it met the following criteria: 'A "Nation, Nationality or People"... is a group of people who have or share a large measure of a common culture or similar customs, mutual intelligibility of language, belief in a common or related identities, a common psychological make-up, and who inhabit an identifiable, predominantly contiguous territory' (Federal Democratic Republic of Ethiopia 1995: 14, Article 39: 5).

local nationalist groups had played in overthrowing the previous regime, and because of the negative experiences that non-Amhara groups had under the strong central states (Pausewang 1994; Young 1996; Joireman 1997; Clapham 2002). Clapham (2002) also claims that the idea for structuring the country along ethnic lines was borrowed directly from Stalin's work on nationalities in Russia. But this model was also thought appropriate for more positive and country-specific reasons (Young 1996). Ethnicity, or tribalism, has been the scourge of governments in African countries. By building ethnicity into the state structure and giving each ethnic group the right to self-administrate, it was hoped that the wider state could avoid becoming subject to inter-ethnic struggles.

There were other ways in which those who were marginalized under previous regimes, particularly minority groups, were set to benefit from these policies: by valuing the local culture it was intended to reverse the impact of years in which non-Amharic groups were considered inferior. By using local educated people in government, employment was brought to people in these areas. The local bureaucrats knew the area, were aware of the needs of the people and spoke the local language. It was thus hoped to reduce the gap between government officials and local people, a gap that had been significant and problematic in previous regimes (Pausewang 1994). By giving authority to local people in local government, it was hoped that a synergy could be produced between the aims and policies of government and the local culture, leading ultimately to more effective, culturally appropriate and sustainable policies. The state's representatives and policies were to be chosen and ratified through democratic elections at all scales, from the local village level to the regional and then to the national federal government level.

These developments can thus be analysed in terms of the local conditions and history, but they also correspond to trends and developments at an international level, as federalism is an important feature of the post-cold war world (Smith 1996). Currently there are wider global movements that stress the legitimacy of local culture and the value of fostering local ethnic diversity. These approaches value indigenous knowledge and capacities, aspects that, in the main, were previously ignored in the face of the uncritical promotion of scientific or modern rationality (Blunt and Warren 1996; Leach and Mearns 1996). The majority of development approaches that promote indigenous forms of organization, knowledge and practice have been associated with social movements, non-governmental organizations (NGOs) and civil society in general. The Ethiopian case is particular because here it is being promoted by and is integral to the main structures of the state.

There are dangers within this state structure. Ethnicity is not fixed or 'natural'; it is constructed over time and within a particular political context (Ranger 1983; Spear 1993; Turton 1997). Turton (1997) argues that the extent to which ethnicity, as a form of communal identity and a sense of belonging, comes into 'salience' depends on the extent to which it is used and stressed by the wider society. It is open to manipulation and exploitation, and 'has enormous potential to mobilize and motivate collective behaviour' (Turton 1997: 78), even resulting in violence against 'other' groups. Although in Ethiopia ethnic groups have been equated with 'nations', the extent to which a local ethnic group mobilizes itself as a group and is nationalistic varies. It may depend on the way in which ideas of self and of culture and identity are used in the process. To what extent does ethnic decentralization reify or exaggerate forms of ethnic identity? Writers such as Krylow (1994)[4] and Joireman (1997) suggest that the policy can exacerbate differences and lead to deteriorating relations between neighbouring groups.

The Emergence and Nature of Konso Special Wereda[5]

Konso Special Wereda was formed in 1993 and at that time comprised the main Konso homeland and a smaller area inhabited by Gawwada people. Later, part of Gommayde was added; previously occupied by northern farmers, Konso, Burji, Dirashe (from Gidole), Amarro and other people have now settled there. In 1996, Gommayde was divided between its neighbouring special *weredas*: areas settled by Burji went to Burji Special Wereda, by Amarro to Amarro Special Wereda, and so on.[6]

In 2002, I explored with local Konso elites who had been involved in the establishment of Konso Special Wereda the reasons why and how Konso gained this status. I was particularly interested in how Konso was conceptualized as a distinct group – when, as shown above, in practice it was highly heterogeneous – and the role of ideas of Konso culture in the new designation. Also under investigation was the extent to which this process had wider repercussions on the conceptualization and status of Konso 'culture' more generally. The following information is based on interviews with members of different government departments and offices, including the Konso Administrative Council, the Ministry of Agriculture, the Ministry of Culture, the Ministry of Health, the Ministry of Capacity Building, the Konso Judiciary and the Women's Office. Many of the themes that emerged in these discussions were similar. Below the words of one official at the Ministry of Agriculture, who agreed to let our interview be taped, are used to show the nature of the process. Some words of another man from the judiciary are also used as he permitted an in-depth interview in which I could take detailed notes. Both men were heavily involved in the early stages of the establishment of Konso Special Wereda.

At the initial change of government in 1991, Konso was included in north Omo Zone and the zonal headquarters were in Arba Minch. This essentially perpetuated the previous administrative situation, in which Konso was an *awraja*[7] under the administration of Arba Minch, the headquarters of Gamo Gofa Province. In the new government, the educated Konso people were invited to come and participate in the

---

4. See also editor's comments in footnote in Krylow (1994: 231).
5. Some work was carried out on this subject in 1995 and 1996 and preliminary findings were published in Watson (2002). This section draws mainly from fieldwork carried out in 2002, funded by the Nuffield Foundation, which explored the development of the special *wereda* in more detail.
6. This division might have been expected to be extremely contentious. Interviews in the region in 2002 suggested that there is an ongoing tension between Dirashe and Konso over a forested area, but that, apart from this, the Konso at least thought that the division had been rather satisfactory.
7. Unit of governance above the *wereda*, pre-1991.

new administration and to represent the Konso people. This was a great change, as previously the majority of educated people had been excluded from their home areas and had been sent to work in other areas in the south. As the man who now works in the Konso judiciary explained:

> Before we had no such chance [to work in Konso], even they did not allow us to come to our land. For example, I was for more than ten years out in other regions, in Arba Minch and Chencha, where I was a teacher. I was a teacher here before, but I was expelled from here. [Why was this?] Because they didn't like those who can read and write, those who had little education even they didn't like them.[8]

In a separate interview, another man, also now working in the judiciary, described the pre-1991 situation as it was for elites: '[Pre-1991] the administration was by other people. It was not by the children of Konso society. It was by Amhara, Kambata and Gamo people. If a Konso person was working here, these people became afraid; they said that they will tell our secrets, they will control us or they will become corrupt.'

The educated people were few, but nonetheless they decided that Konso would be better if it stood alone under Awasa than being included in the larger north Omo Zone. The man, who is now in the Ministry of Agriculture, Konso, explained:

> [At the time of the Transitional Government] I myself was elected to be a member of the electoral commission, to represent Konso people. At that time the criteria for representation were language and culture – these were the main things. Konso is unique out of northern Omo. The others are not Cushitic, the others are Omotic and their languages are Wolaitinya and Gamunya. Gamu and Dawro and Gofa – all these are similar except for some differences in some words, so that Konso is unique. And Dirashe is also unique, according to these criteria. So I asked them, 'How can we live together, under one political guideline, or administrative guideline?' Then the chairman said, 'Now, go away from us, you are Konso, you can choose your way, but you cannot be included in north Omo.'

He described the way in which Konso was to become a special *wereda*, together with Yem (on the side of Jimma), Dirashe, Burji and Amarro: 'These are just like islands in the region, having their own culture, their own language and their own psychological make-up according to their culture.'

When pressed to describe further the uniqueness of this culture, the language was brought up again, but here similarities as well as differences were emphasized. Konso, it was explained, has 46 per cent of its language in common with the Oromo, a point that is made in Hallpike's 1972 ethnography of Konso. It was interesting that

---

8. The interviews with elites were in English; the interviews with people in the village were in *Afa Xonso*, the Konso language.

the speaker was one of the few people in Konso to have a copy of his book. As well as demonstrating the enduring power of the written word, this raised the question of whether or not it was ever thought that Konso might be allied with Boran, a main Oromo group. The response to this was as follows:

> We have friendship and neighbourhood with the Boran, and also Guji sometimes in the Rift Valley, but because of some cultural actions Boran were killing Konso and Konso were killing Boran in the past. Konso are very polite and good to their neighbours, but when someone comes as a thief, or as a bandit, they are taking a great measure to make war and to kill Boran. So this made it not possible to come together.

Politics also came into the equation, more specifically the fear of being drawn into association with the Oromo Liberation Front, currently fighting against the government. He continued:

> We saw the ambition of some people to join the Oromo [was] because the Oromo are in different lines now. One is the Democratic Oromo and the other is like the Liberation Front. And may be the one who gave us advice to join them, he may be on the side of the Liberation Front. Then people would be into conflict and so we didn't like this and we had to be a special *wereda* and join the southern people's region and administrate ourselves.

Ideas about culture, language and political differences and similarities have played a role in the setting up of Konso Special Wereda, but these were often secondary to more practical and logistical concerns about what could be achieved by this structure and obtained for Konso people. This is the case with the Oromo question also where these cultural and political issues were there, but the practical issues were presented as insurmountable:

> If we were to be joined with Oromo then we have to go to the zone, and the transportation condition it is not good, people can even be killed on the way. If people are to apply to solve their problems they must walk to Yavello, and also to Negele Boran, and it is very far. And people here are not perfect in the language. They cannot read and write the application properly.[9]

All those serving in the new administration who were interviewed stressed that the establishment of the special *wereda* had meant that bureaucratic processes were shortened, and thus they had a closer connection to the resources of the central state. If there is a problem now they go straight to the offices in Awasa, the headquarters of the Southern Region; there is no need to process any business through intermediary zonal offices. They also believed that their ability to gain access to those

---

9. Ministry of Agriculture employee.

resources had improved, as they were no longer in competition with other people in the zone:

> The nature of people is selfish and when I was born in Konso, if we are budgeting and I have all responsibility, I like to take more budget to Konso. The same is that to northern Omo: some people are to Boreda, some people are to Chencha, to Dita *wereda*. They are making budgets even for development projects like schools, clinics. Then those people who are now in position are now inclining to their native places. So that other people are suffering.[10]

In all interviews with elites, emphasis was placed on the way in which becoming a special *wereda* gave them access to jobs in local government. In some cases, for example court officials, those employed in the new special *wereda* had the same status as zonal officials, and so received higher salaries. A great advantage of the employment of 'Konso children' in government offices, courts and health stations is that they can speak directly to their clients. A member of the Konso judiciary explained this (where by 'own children' he is referring to children of Konso in general, rather than of a particular family):

> Now if someone goes to the administration or the police or the soldiers he gets his own children. The same is the case in the agricultural office and the education office. There is no language problem, a person can explain his problem to his child. Before he is very fearful. He thinks if he goes and speaks his mind, he could be even put in prison or fined.

The new special *wereda* aimed to secure resources: they were concerned with the business of nation-building in terms of schools and health stations, rather than with national identity. The majority of administrators listed the achievements of the new special *wereda*: an increase in the number of primary and secondary schools since 1991; an increase in the number of Konso people who have received primary, secondary and further education; an increase in the number of health stations that have been built, many in remote areas and staffed by local people; and an increase in the number of Konso people who are now in employment.

In contrast, there seemed to be little interest in developing further the ideas, understandings and status of the Konso culture or language. In 1995, I attended a workshop on writing the language, where it became clear that this was going to be a difficult task (Watson 2002). In 2002, this work had completely stagnated. Very few resources and personnel had been applied to the task. The majority of officials believed this work to be the responsibility of the Department of Culture, but the Department of Culture seemed confused about its responsibilities. In the years since 1991 it had existed in various forms: for instance, the Department of Culture and

---

10. Ministry of Agriculture employee.

Sports and the Department of Culture, Information and Tourism. In 2002 it was divided into two branches: Culture, Sports and Youth, and News and Information. Despite interest in the local culture, no progress had been made in documenting any aspects of local culture or history. Any effort in this direction was hampered by lack of expertise and what were considered more immediate pressing concerns. In 2002, the Department's main concern was taking tourists to the villages for a fee that contributed to the *wereda*'s main budget. They had some information on the local culture, but the majority of materials had been written by outside researchers who had sent their findings to the office.

In summary, the establishment of Konso Special Wereda was seen mainly as a way in which Konso had increased its share of resources and had a government that was closer to the people. Pre-1991, the Konso elites had shared the same experience of elites from other areas: they had been marginalized from their home areas and from a share in the resources and responsibilities of government. Unlike the other areas, however, where this marginalization led to the formation of resistance movements based around ethnic and nationalist identities, in Konso the number of elites was few, and pre-1991 they cannot be considered to have developed into a nationalist movement. Post-1991, the local elite have grasped the opportunity to govern themselves, seeing advantages for themselves and their people in this structure, but the extent to which they exist as a nationalist movement today remains questionable. The educated elite have not developed an ideology to bolster their claims to be a special *wereda*. Konso culture has not been celebrated or emphasized as part of this process at an official level. Instead, it has been given a back seat, as modern developments have been promoted in terms of schools, education, voting participation and so on. Although these developments are not necessarily in opposition to the local culture, they have often been interpreted in this way.

One possible reason for a lack of interest in local culture and the opposition drawn in minds between development and local culture is the importance of some Protestant Christian movements in the area. I put this argument forward in 2002, and during fieldwork this was confirmed by the official from the Ministry of Agriculture whose interview I have drawn on above.[11] When I asked him how the establishment of the special *wereda* had changed the status of Konso culture, he replied:

> Consciousness limits this. For example in Konso we do not have many educated people. Also the religion also affects this. As the church is assuming that everything to do with the culture is bad, bad, bad, bad, bad. But after they will be well matured they will be very sorry, and they will go looking for it and they will not get it.

However, among Protestant Christians there is no unity at present. Many different Protestant Christian denominations have come to Konso in the past five years, and some churches within the same denomination have split and have been fighting

---

11. Many of the sentiments he expressed were confirmed by other interviewees.

among themselves. This has led to a lack of authority on the part of the church and its effect is not as strong as it has been in the past.

Overall, therefore, the elites in power in Konso have spent most of their energy in addressing the problems and needs of modern development and practical nation-building. Work on the culture and the language has been seen as secondary when the health care, education and food shortages are demanding attention. The priorities have been to secure livelihoods and to turn to questions of culture later. But this has made the administration vulnerable as others can use this lack of attention to culture and language to their own advantage. When I visited in 2002, there was the new presence (however marginalized) of a political opposition party that had stood in elections in 2000 against those who had previously enjoyed relatively unquestioned power. The Konso People's Democratic Organization (KPDO), known in Konso as *Kohedit*, were aligned to the governing EPRDF and had been the main ruling party in Konso. The Konso People's Democratic Union (KPDU), known in Konso as *Kohedih*, were allied with '*debub hiberit*', the 'southern union' opposition party, and formed the new opposition.

When I met a member of the opposition party in 2002, who had stood in the elections in 2000, he asked me if I knew what his electoral symbol had been.[12] 'The drum', he told me; he stressed that the drum was the symbol par excellence for Konso people's respect for truth, as the *apa timba* (father of the drum) is the customary institution that stands for truth, order and culture. He told me that, because of his symbol and because of his campaign, the administration in power in the *wereda* had been forced to call a meeting with the people and proclaim that 'they [the administration] are not against Konso culture'. The need to call a meeting was put forward by the opposition as evidence that, far from promoting Konso culture, the administration of the time were against it and were not truly representing the people. It is difficult to verify this story at present, but it demonstrates how the issue of the role of culture becomes manipulated in political debates, and also the political capital that can be gained from using Konso culture as part of the political argument.

This fits with the view that Konso administration has thought little about Konso culture since they came to power. Beyond the occasional political rally at which there is Konso dancing or attendance at a key ritual, there has been little engagement with cultural issues. The culture and language of Konso were a concern of importance when the division of the states was first established along cultural lines. Following this, there was little motivation for the administration to think about culture in political terms or to invest in the development of the language or studies of the culture or history. There were people who were non-Konso in the administration, and some of these have complained that, because they are not Konso, they have not had the opportunities for promotion that others have had. But, among the majority who are Konso, their domination in the new special *wereda* has been so strong that it has not seemed necessary to engage in the questions of 'Who and what are the Konso?' In the main there have not been boundary struggles, or threats from other groups. The elites have only had to start thinking about the definition of Konso

---

12. This symbol is used to represent the candidate on voting papers.

again when forced to do so by the opposition party, who themselves may be attempting to call on references to local culture as a means to generate support.

This lack of development of ideas of what constitutes Konso may have advantages: it prevents certain aspects of Konso culture being officialized and prioritized over others, and also prevents ideas of Konso becoming essentialized or hijacked for political reasons. The elites in Konso also appear to be aware of the dangers inherent in ethno-nationalist projects: in discussion several elites expressed reservations about the development of a strong official model of Konso culture on which the identity of Konso Special Wereda could be based. This was not only because many felt that the cultural ideas conflicted with other ideas of themselves (as formally educated, often Christian, people). But it was also because they realized the harm that would be done by developing a strong model of Konso culture, in terms of the way in which it would prioritize certain aspects of the culture (and people) and exclude others – both within Konso and outside. This lack of a strong ideological base for the new special *wereda* may, however, also have led to a lack of unity of purpose among the educated elite: the administration has been characterized by internal disputes, accusations of corruption, changes of personnel, demotions, arrests and internecine strife.

View from the Village

This lack of discussion of what or who are Konso was not equated with a lack of interest in 'cultural matters' at a village level. In at least one village, Buso, in the central area of Konso, where I have worked over a period of seven years, the new political context in Ethiopia has combined with certain other factors to generate a resurgence in local cultural practices and ideas.

In order to understand these developments, it is important to look at the historical background: in Buso village, the events that took place 1974–91 were similar to the pattern of events that have been described for other areas of southern Ethiopia (Donham 1999). The modernizing efforts of the Derg political cadres dovetailed with the anti-tradition elements of Protestant Christian movements to undermine local customary office-holders. Following this, ritual leaders rarely carried out public rituals because their spiritual position had been attacked and their wealth had been undermined. They found it difficult to raise the necessary capital for the rituals (livestock for slaughter and grain for brewing) (Watson 1998). Those who hold customary positions (I refer to these as customary office-holders) to whom were attributed supernatural powers such as prophecy and healing were paraded in marketplaces and were publicly humiliated. Other institutions were also attacked, including the 'father of the drum', *apa timba*. The drum, the symbol of authority and truth in the village, circled between a given set of families. Those families who held the drum were *apa timba* and, for the time they held the drum, were responsible for fining and punishing those who lied or did wrong. In many areas the drums were captured and confiscated during the Derg years. In Buso, the drum was hidden and no longer passed between the families; the position of *apa timba* became defunct.

In 1995/96, when I carried out fieldwork in this village, there was talk of restarting the cycle of the drum. After an initial attempt it stalled, as in one year the man who was due to accept the drum refused it. One of the reasons given was that many of his family were Protestant Christians and had put pressure on him not to accept it as the responsibilities included a sacred, but non-Christian, dimension (Watson 2002). In 1995/96 Protestant Christianity was very strong in the village and the church had a regular congregation of more than 500 people.[13]

Returning in 2002, I found the drum was circling again and its holder actively receiving, judging and fining cases. Moreover, I found that the current *apa timba* and several other people who had held the drum or who had other positions of customary authority had been elected onto the new state organizational body, responsible for the administration of the village. This body had fifteen representatives elected by each of four village wards (*kanta* in the local language[14]) making a governing body for the village of sixty people. The body was described in the local language as the *parlamma* (sometimes *parmalla*) (see Plate 9.2). From within these sixty people, a smaller executive body was selected, a 'cabinet' (sometimes this word was used locally), who essentially carried out the same functions as the old Peasants' Association committee. The same organizational structure of 'parliament' and 'cabinet' existed at village, *wereda*, regional and national level, and the village also had three elected representatives on the *wereda*-level '*parlamma*'.

The customary office-holders who became members of the *parlamma*, including the current *apa timba*, were elected in the same way as the other members of the *parlamma* (the *parlamma* included fifty-one men and nine women, all of whom had been elected by the wider village). The village did not make a conscious decision that the customary office-holders should be represented on the *parlamma*, but I was told that the individual office-holders were elected because they were considered by the wider village to be respectable people who had the capacity to lead. However, the fact that these customary office-holders were represented and that they concurrently served the village in their customary capacity and in a formal state capacity represented a significant shift in the relationship between the people, the customary authority and the state. Not only had customary office-holders been reinstated and reinvented, but they were playing a key role in the new statutory administrative structures.

One possibility here is that the state was co-opting customary leaders as a way to control the people more effectively, as a kind of indirect rule. I do not believe that this is the case here, partly because the customary leaders were not appointed by the state. In 2001, a formal election took place. Those interviewed stressed that the *parlamma* was the result of the new democratic processes, but the democratic election that took place in the village, although democratic in the minds of the people, did not live up to accepted democratic ideals.[15] Prior to this formal election,

---

13. The Christians were mostly part of the Ethiopian Evangelical Church Mekana Yesus.
14. There are actually five *kanta* in this village; two were amalgamated for this process.
15. These democratic procedures may have been the first time a large-scale election of this level of formality had taken place with ballot papers, but it was not the first experience of the people with the idea of elections. Several elections had taken place during the Derg period (for the PA committee) and previously during this regime.

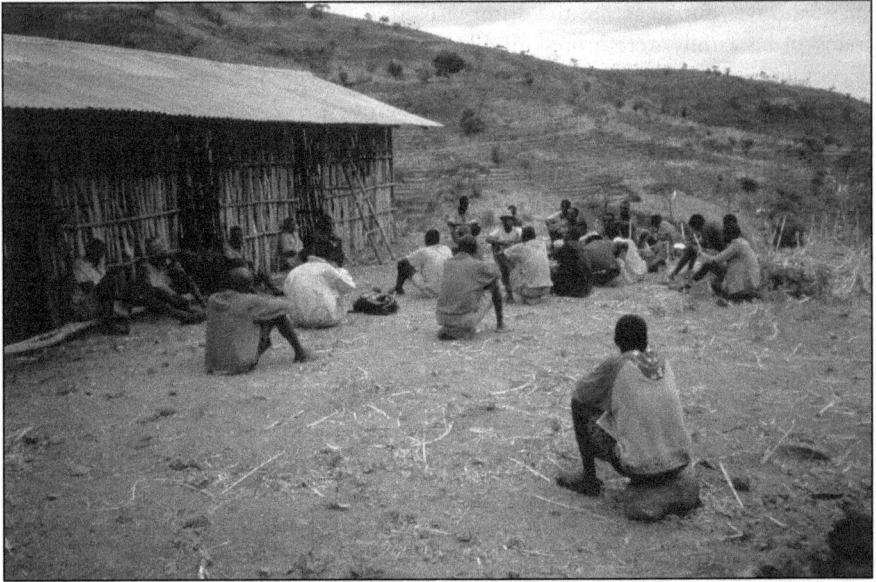

Plate 9.2 Village *parlamma* meeting (photo: E. Watson)

the village held public meetings for each group of people that represented an electoral ward. There they selected, through discussion, the people who were to stand for the election. As they explained: 'They make a meeting and then the people will discuss the different people. They will say, "He will do this in a good way; he will manage the village in a good way." Thus, the ward will tell a person's history; others will also tell his bad history. In this way they will choose the candidates.'[16]

In the election itself, sixty people stood, and sixty people were elected. This chapter is not the place to discuss the faults with this democratic process or to evaluate the argument that the Konso in this village subverted the imposed democratic procedures and replaced them with their own. What is important here is that, through becoming elected to the *parlamma*, customary authorities (such as the 'father of the drum' and ritual and clan leaders) formed an alliance with the structures of authority of the state. Some of these customary office-holders lost power and control over property and some lost their cultural legitimacy during the Derg years. Becoming a member of the *parlamma* has given them a new source of authority and a new position, which they have combined with their old responsibilities.

When asked to comment on what has happened, friends in the village said, 'Now the culture (*aatta*) is strong.' Several people asked me rhetorically, 'How can we live without a drum?' I found this question surprising, as in the mid-1990s they had lived without a drum seemingly quite effectively. Previously, 'culture' in the village was something that was lived and breathed, practised and performed and

16. Group discussion with *parlamma* members, 2002.

taken for granted. But now there was also an awareness of Konso culture as something that it was desirable to reinstate, to protect and to cultivate. This represented a new consciousness of Konso culture that has resulted from their historical experience.

Four factors can be identified as contributing to the resurgence in 'culture' in this village. First, the way in which cultural office-holders were attacked and suppressed during the Derg regime meant that they were not tainted by the problems and abuses of power that were associated with that period. In 2002, there was still some optimism about the opportunities presented by the new regime, and the customary authorities seemed to present a fresh source of leadership talent that had been ignored in previous years.

The second contextual factor was the decline in the force of Protestant Christianity in shaping ideas about the legitimacy of cultural beliefs and practices. In 1995 there were approximately 500 congregation members in a village of about 2,500 people. For many of these Protestant Christians, customary practices were seen as dangerous, even equated explicitly with devil worship (Watson 1998). The influence of Protestant Christian ideas in 1995/96 influenced more than just their converts. The Protestant Christian church, the Ethiopian Evangelical Church Mekana Yesus (EECMY), was evangelizing in the villages, and was also a source of health care, food relief, education and employment, and it was highly respected.

In 2002 the situation in Konso had changed considerably. EECMY was already subject to internal dispute in the mid-1990s, but later the disagreements became entrenched and divisive. The EECMY community has split into two churches: a church mainly populated by younger people, who prefer more lively music and liturgy, located near the marketplace; and a church on the mission compound mainly populated by older people, who prefer more sedentary and traditional styles of music and worship. The older community accuse the younger community of charismatic forms of worship which they say are outside respected church practices. Some claim that their singing and dancing is not worship of God, but a vice. In retaliation, the younger community accuse the older community of corruption. They claim that the pastors preach one thing and that in their personal lives they are doing the opposite. They also assert that they distort the material in religious texts to propagate their personal moral views. This disagreement has boiled over into physical conflict on more than one occasion.

While this conflict has continued, enthusiasm for this church has waned in the village. Evangelizers have not been visiting as they were before. Those who were Protestant Christians ask, 'Why should we follow people who tell you not to fight, but who are fighting all the time?' In addition, there are now many new Protestant Christian denominations in Konso, and the competition between them has made people sceptical about one individual church's claim to the truth. With the belief in Protestant Christianity came the promise of a better life, spiritually and materially. As this has failed to materialize, some disillusioned people have switched from being Protestant Christians to what they refer to as the 'culture'.

Thirdly, the rhetoric and policy of the wider state that celebrate local culture have undoubtedly contributed to this resurgence of culture. Locally, this context is understood through the idea of democracy, which in turn is understood as having the right to live as one chooses. More specifically, under the previous regimes, minority groups and cultures were considered inferior; now everyone has to be respected, whoever they are and whatever their culture. The chairperson of Buso village explained:

'Before the courts looked down on people. They didn't take a person to be a person. Now, no one can just hit you – this is democracy. Now, everyone is equal. Everyone is one. You are a northerner, you are foreign, but you can't be considered more than me.'[17]

The insistence on preserving the unity of the village was a fourth factor in promoting cultural practices. The village embraced democracy but it rejected the idea and practice of multiparty democracy. It was thought that if a village had unity then it would be able to achieve many things. This fitted with older indigenous ideas, in which agreement and unity are thought important; quarrelling and disputes are thought to lead to famine and sickness. With unity it was thought that the people would be able to solve many of the problems that faced them day to day; it could send children to school and carry out public works programmes, such as cleaning the pathways or constructing fences and public buildings. Promoting those with customary authority in the *parlamma* was seen as one way in which a degree of unity could be achieved. It also created a powerful alliance between the authority vested in the state and the authority vested in tradition. In comparison with the time when I previously lived in the village, the people did appear to be more united and able to achieve more because of this.

This outcome was not the same in all villages. There are villages where there is no unity and there has been no conscious promotion of cultural practices or institutions. Instead of unity in these places, since 1991 there have been disagreement and dispute that have led to protracted conflict among village inhabitants. For example, in Gamole village, less than 5 km away from Buso as the crow flies, there have been division and fighting. This has been viewed by the people of Buso as a cautionary tale and another reason why the people of Buso should unite together.

What accounts for the difference? In Gamole, the customary office-holders fared better during the Derg regime, as there was at that time in this village something of an alliance between the customary authorities and the Derg committee. Following the change of government, the customary authorities did not exist as a fresh source of cultural and political capital or of leadership. Whereas in Buso Protestant Christianity has waned, in Gamole there is a strong growth of Orthodox Christianity, which provides alternative forms of organization, inspiration and authority to rival the customary. Finally, the context of democracy and 'permitting difference' has not led to a revival in customary practices and the position of office-holders; instead it has led people to say 'Everyone can now do what they want' and

---

17. Interview in Buso, 2002. The interview was in *Afa Xonso*, but the interviewee used the English word 'democracy', which has entered the language.

'Nobody has to obey anyone.' The chairperson of Gamole village explained (using the word 'democracy', which has entered the language): 'You know this democracy? In this democracy everyone has rights. But this leads these people not to obey ... Since it is democracy everyone can follow his own road. Even I can go to Addis Ababa and live there.'[18] Gamole is a village with many different social groups. The differences may have been exacerbated by the opportunities provided by democratic procedures. The customary office-holders have been discredited by their cooperation with the previous regime, and no one is able to unite or to make peace in the village.

Conclusion

In the context of ethnic decentralization there has been a rhetoric about the celebration of local culture and history, but in practice this has been little developed by the elites who are engaged in the project of building and maintaining Konso Special Wereda as an administrative unit. Origin myths, traditions, ideas of culture and language are often exploited as part of ethno-nationalist projects, as a way to motivate and mobilize collective behaviour in order to gain access to resources in competition with others (Turton 1997). In Konso, the status of special *wereda* seems to have provided a structure in which access to state resources has been somewhat improved, and there has been little need or desire to engage in further ethno-nationalism. There has also, perhaps quite rightly, been a scepticism about the ethno-nationalist project on the part of the elites. However, the opposition to the current administration may try to draw on cultural ideas as a way to generate political capital. In addition, if the current administration is threatened, internally or externally, it may also engage more in debates about the importance of Konso cultural practices and language, in order to generate support.[19]

In some villages, there has been increasing debate about the status and role of culture, or *aatta*. Where this has happened, it can be seen as partly a result of the failure of other projects, such as that of Protestant Christianity and socialism, which promised alternative lifestyles to the people, but failed to live up to the expectations they had generated. In addition, the people in Buso at least, who had, as they put it 'gone back to the culture', were of the opinion that 'culture' had the capacity to unite the people. The Konso have experienced tumultuous changes in the past, and continue to struggle with changes in the present day. The traditional cultural ideas and practices, when reasserted or reinvented, have the capacity to provide a comforting sense of self and identity. The difference that exists between what has happened in two villages, only 5 km apart, however, shows the degree of variation that must be taken into account over short distances. The processes of identification are influenced by old and new ideas, reinvented and combined in novel ways, producing many different combinations on the ground.

---

18. Interview in Gamole, 2002.
19. At the time of writing there is discussion of an administrative reorganization that might change the status of Konso Special Wereda again. It has yet to be seen what will emerge from these discussions and the impact it will have on Konso people.

In much of the work about ethnic identity, there is an emphasis on the way in which ideas about the identity and cultural practices of a particular group become particularly salient in the border areas, where one group comes into contact with another. In the multi-ethnic market town of Gommayde, Konso people there have recently constructed a traditional-style meeting courtyard, together with a traditional-style men's sleeping house. This construction symbolically emphasizes their Konsoness and is a reference to Konso cultural practices. It is situated by the side of the road and all who are passing can take note: a classic example of emphasizing cultural identity in the face of difference.

This assertion of identity might have been expected in a border area. But in the centre of Konso, at the village level in Buso, it is possible that the stress on cultural practices and ideas that has taken place recently has also been partly a result of encounters with difference. The Protestant Christian and socialist movements and modernizing state projects have also represented alternative ways of being connected, conceptually and materially, with different places: with other towns, the wider state, Awasa, Addis Ababa and foreign lands. In the past, ideas of similarity and difference would probably have related more to one's ethnic neighbours, and ideas of place in the world would have been constructed through a local geography in which these ethnic neighbours featured more highly. Now, ideas about similarity and difference are mapped onto a wider geography connected with the historical experience of processes in this wider world. As the alternatives have been tried and been discredited, the local cultural forms have been reasserted, although in different ways. The background to this reassertion has been the wider political context, which has provided conceptual and legal 'space' for the celebration of previously degraded forms of local culture.

Chapter 10

Changing Alliances of Guji-Oromo and their Neighbours: State Policies and Local Factors

Taddesse Berisso

Introduction

The Guji are one of the many branches of the Oromo people. In the administrative division of the current regime, they live predominantly in Guji and Borana zones of the Oromiyaa Regional State, southern Ethiopia. The Borana, Arsi, Sidama, Gedeo, Burji, Konso, Wolaita, Koyra, Gamo and Garre are some of the major neighbouring groups of the Guji. Historically, the relationship of the Guji with all their neighbouring groups (except with the Gedeo and the Sidama who live in Wondo Genet area) was characterized by conflict and hostility. Warfare was endemic, generally taking the form of raids against the various Oromo and non-Oromo ethnic groups bordering Guji territory (Hinnant 1977). Conflicts and raids have continued to the present, still occurring sporadically.

The Guji maintained hostile relations with both Oromo and non-Oromo groups in the area, but they considered the Oromo groups (i.e. the Borana and Arsi) as their main enemies. Recently, however, there has been a remarkable transformation in the relationship between the Guji and their Oromo neighbours. The transformation has been brought by changes in the national political system, by an increase in the level of political consciousness and by socio-economic changes. With these developments, some of the groups who were former enemies have become allies, while some age-old allies have become enemies. Based on recently conducted fieldwork, this chapter examines developments in inter-ethnic relations and the formation of new alliances in the study area.

The Guji-Oromo: an Overview

The Guji-Oromo (also known as Jam Jam or Jam Jamtu by some of their neighbours) speak *Afan* Oromo (*Oromiffa*), one of the most widely spoken languages in Ethiopia, belonging to the Eastern Cushitic language family. They have a population of more than one million people. The Guji were conquered and incorporated into the Ethiopian Empire in 1896 by the forces of Menelik II

(1889–1913). Outside the Oromiyaa Regional State, small pockets of the Guji population currently live in the Wondo Genet area and in the Nechsar National Park. The Guji in Wondo Genet area are fairly assimilated to the Sidama culture, as they have lived there among the Sidama for centuries.

Unlike some other Oromo groups that constitute a single section, the Guji are a confederation of three independent, but closely related, groups known as Uraga, Mati and Hoku. Historically, each section had its own territorial boundary and political leader in the form of *abba gada*, 'an age-grade leader'. The three groups are interdependent: they regard each other as blood relatives, act together in case of war against neighbouring groups, help each other during economic crises and conduct *gada* rituals together. Intermarriage is fairly common, and there are few cultural differences between them. Individuals or families from one group were free to move and settle in another's territory.

In the past, the socio-political organization of Guji society was dominated by a moiety-clan-lineage-family structure and by the *gada* system, with a *qallu*, 'supreme hereditary religious leader', at the apex. There are two non-exogamous moieties, known as Kontoma and Darimu, which cut across the three groups. Under these moieties, there are seven non-totemic and exogamous clans each in Uraga and Hoku and three in Mati. Each clan is divided into a variable number of segments called *mana*, literally 'house', which in turn are divided into a great number of patrilineages. The Guji family is an extended patriarchal family. Marriage is based, in most cases, on self-selection and on arrangement between the families of the bride and the groom. Polygyny, primogeniture, patrilocal residence and levirate are some of the other features of the Guji marriage and family patterns.

The *gada* is an age-grade system that divides the stages of life of individuals, from childhood to old age, into a series of formal steps. There are thirteen such steps in contemporary Guji society. Transition ceremonies mark the passage from one stage to the next. Within each stage, activities and social roles are formally defined, in terms of both what is permitted and what is forbidden. The ideal length of time in one rank is eight years. In the past, the *gada* system assumed military, economic, legal and arbitrational responsibilities. Some of these *gada* responsibilities have been gradually eroded due to the conquest and incorporation, introduction of new world religions, internal dynamics of the *gada* system itself and other factors.

The Guji have a mixed economy of animal husbandry and crop cultivation on fertile land that stretches over a wide variety of altitudes. They subsist mainly by growing grains such as barley, maize and teff (*Eragrostis tef*), pulses and enset (*Muse ensete*). But their real wealth consists of cattle, sheep, goats and horses. Emotions and pride are centred in stock. People who do not own cattle are not considered to be 'proper' Guji (Baxter 1991: 9). Cattle are important not only for economic purposes, but also for social and ritual life. The social status of a person among the Guji finds its expression in the number of cattle owned. The owner of many head of cattle is a respected person. Cattle are used ritually for sacrificial purposes.

With regard to religion, Guji have developed a very complex set of beliefs and practices. They believe in *Waqa*, 'God', and in the existence of *durissa*, 'devil'. They have *Woyyu*, 'sacred shrines', at which prayers and sacrifices are made to *Waqa*. They also believe that ritual power is vested in certain individuals and families: for example, the *qallu*, 'supreme religious leader', and *abba gada* are considered to be *Woyyu*, 'holy', and *Worra Kalacha*, the 'virile family', respectively. Attached to these positions are responsibilities for rituals and religious practices, including communication with oracles, spirit possession, complex and varied divinations and other customary beliefs. With the introduction of the great world religions and modernization, Guji customary beliefs have been changing fast. There has been mass conversion of Guji farmers to Christianity (Protestantism in particular) and to Islam, especially following the 1974 Ethiopian Socialist Revolution (Taddesse 1995).

The Guji consider their homeland to be the very ancestral cradle of Oromo culture. Indeed, from analysis of historical linguistics, oral traditions and cultural data, many scholars have now concluded that the Oromo originated in and around the areas currently inhabited by the Guji, Borana and Arsi Oromo (Haberland 1963; Lewis 1966; Asmarom 1973).[1] It is from these areas that the Oromo launched their vast expansion in the sixteenth and seventeenth centuries, organized in part on the basis of the *gada* system. It is not clear, however, when or why the Guji separated from other Oromo groups and formed an independent territorial unit.

Guji-Oromo Relations with their Neighbouring Groups

The Guji and their Oromo neighbours

The Borana and the Arsi are the two major Oromo neighbours of the Guji. The three Oromo groups (i.e. the Guji, Borana and Arsi) are united by common language, custom, way of life, historical experience, a common world view underpinned by the *gada* system and customary Oromo religion. However, the relationship between these Oromo groups has long been characterized by hostility. They fought each other for economic, political, cultural and structural reasons (Duba 1986; Taddesse 1988; Tedecha 1988). The uncordial relations were also exacerbated by prolonged negative attitudes towards each other (Zerihun 1999).

Guji oral tradition attributes these hostile relations to an incident that happened among the founders of the three groups. According to the tradition, the founders of the three groups (Gujo, Boro and Arse) were half-brothers. One day, it was said, their mothers quarrelled and fought, and their respective sons (with other members of the family) joined the fight in support of their mothers. In the fighting, people were killed on all sides by their half-brothers. The mothers, consequently, left their areas of residence and joined their fathers' homes with their remaining sons. Since then enmity has existed between the three groups; the killing of each other has been seen as the legitimate avenging of the spilling of their brothers' blood.

Whatever the causes of their conflicts, the three Oromo groups considered each other as *Siddi Saddin*, 'three enemies against one another', despite their common origin and culture. Within the framework of *Siddi Saddin*, Guji, Borana and Arsi all consider each other as '*akaku*', 'human beings equivalent to each

---

1. Recently, however, it has been argued that the Oromo cradle-land (with other Cushitic groups) was far north, in the present Ethio-Sudan regions (see, for example, Alemayehu et al. 2004).

other'. The killing of *akaku* awards the killer with the highest honour and prestige, known as *mida* or *mirga*.

Mida/Mirga is considered a heroic act of bravery. It relates to the killing or capturing of a man from among the *Siddi Saddin* or to the killing of four types of dangerous wild animals: elephant, lion, buffalo, or rhinoceros. The killer or captor gains great social recognition. *Mirga* qualifies a person to smear his hair with butter for two years and entitles him with the right to sing the song of killers (*gerarsa* and *dadu*) and to participate in the ceremony of killers (the *kuda* festival; see Taddesse 1988; Tedecha 1988).

To the three Oromo groups, the killing or capturing of a human being from any other group does not qualify as *mida/mirga*.[2] By the same token, the Guji call members of the *Siddi Saddin* male and all others female.

The concept of *Siddi Saddin* does not deny the fact that these three Oromo groups are the same people. They believe that they originated from a common ancestor. But their fighting suggests that they have found each other to represent more dangerous and threatening rivals than other groups. They say '*tirun gnapte: gnapa calitte*', 'blood relatives who become enemies are more dangerous to each other than casual enemies'. Thus, the *Siddi Saddin* represents the defence of oneself from the more dangerous enemy of one's own blood (Tedecha 1988).

The Guji and their non-Oromo neighbours

Since their emergence as an independent territorial unit, the Guji have lived in a condition of open hostility with some of their non-Oromo neighbours and peacefully with others. According to informants, before Menelik's conquest of the south in the last quarter of the nineteenth century, inter-ethnic contacts and relations were greatly limited to the border areas. The barter trade that existed among some of these groups (including the Guji) was not developed to the point where it could create significant relations and cooperation among them. Intermarriage was not encouraged among most of these groups.

Even after Menelik's conquest, the relationships between most of these groups were limited, to a large extent, to marketplaces and towns; these are neutral zones where different ethnic groups meet and exchange their produce and information about their social and natural environments. Apart from marketplaces and towns, contacts were limited and the passage into each other's territory was often dangerous. The causes of and motives for the conflicts were economic, political and structural (e.g. *gada* requirements) and cultural.

Among the non-Oromo groups, the Guji's relations with the Gedeo and with the Sidama of Wondo Genet area differed from the usual pattern. Guji relations with the Gedeo were not always harmonious, but they were much better than, for example, those with the Arsi, the Borana and the Sidama in the south. These better relations were probably the result of two factors. First, the Guji and the Gedeo

---

2.   This has been changed over time and other groups, such as the Sidama and Wolaita, are now also included as *mirga* subjects.

consider themselves blood relations. In one of the Guji origin myths, it is said that Darasso (the founding father of the Gedeo) was the elder brother of Gujo (the founding father of the Guji). Both of them were said to be the sons of Ana Sora. Consequently, even today Guji do not drink milk, *dadi boka* (honey wine used for ceremonies) or give blessings at ceremonies until the Gedeo have received the first share, the symbol of respect usually given to elder brothers.

The second and probably the most important reason was the economic interdependence between the two groups. The Guji obtained the cereals and *workil qocho* they needed from the Gedeo, while the Gedeo in their turn obtained cattle and cattle products from the Guji. There were also intermarriages between the two groups and, when necessary, they fought in alliance against their common enemy, the Sidama. The Guji and Gedeo avoided killing each other. If such an event occurred, the killer would not wear butter on his hair, sing *gerarsa* (killer's songs) or organize a *kuda* festival. In 1998, however, the Guji and Gedeo fought a bloody war on issues related to territorial delineation, which consumed the lives of more than seven hundred people from both sides. There was also the destruction of innumerable amounts of property.

The Guji-Sidama relationship in the Wondo Genet area was different from the relationship between the two groups further south. There, bitter hostility and war characterized their relations. Economic factors, mainly the southward expansion of the Sidama into the domain of the Guji in search of resources, were cited by informants as causes of the hostile relationship.

In Wondo Genet, however, the relationship between the Guji and Sidama has been peaceful and based on mutual benefits from the time of their initial contact. It is believed locally that the contact between the two groups started with the northward expansion of the Sidama to the Wondo Genet area from the highlands. They developed reciprocal relations in which the agricultural Sidama supplied the Guji with agricultural products, mainly *workil qocho*, while receiving dairy products from the pastoral Guji. Besides this, the Guji in Wondo Genet were cut off from the other major Guji groups living in the south, and thus were barely affected by the nature of relations between the two groups outside Wondo Genet. Moreover, the demographic and military supremacy of the Sidama in Wondo Genet and their mutual hostility to the Arsi promoted the development of cooperation and a cordial relationship between the Sidama and Guji (Zerihun 1999).

Increasing interactions resulted in the cultural integration and assimilation of the Guji with the Sidama. In some Peasant Associations (PAs), such as Yubo, Arema and Edi, in the southern and south-western part of Wondo Genet, the Guji are highly assimilated into the Sidama culture and some of them speak only the Sidama language (Zerihun 1999). The high degree of interaction also allowed extensive marriage alliances between them. In Guji oral tradition in the Wondo Genet area, the positive relations between the Sidama and Guji are attributed to the genealogical relationship the Guji have with one of the Sidama clans, Malge, and the marriage alliance with the remaining Sidama clans. As with the Gedeo-Guji relations, the Sidama-Guji relations were cordial and strengthened by their alliance against their common enemy, the Arsi, until 1991. However, in the post-1991 period, the cordial relationship has been undermined by the changes in the national political system.

Changes in Inter-ethnic Relations and in Ethnic Alliances

In Ethiopia, the imperial and the Derg governments adopted similar policies on the question of ethnic identities and ethnic nationalism. Both governments promoted a centralized policy in which no recognition was given to ethnic identities. Ethnic and linguistic boundaries were barely used to organize administrative regions. Contrary to this, the Ethiopian People's Revolutionary Democratic Front (EPRDF)-led government instigated a radical policy change as soon as it came to power in May 1991. The new policy was based on federalism, and federal regions were organized along ethnic and linguistic boundaries. A policy of self-government was implemented, at least in theory, in all the newly established regions, where local people themselves were given local political and administrative posts. Ethnicity-based political organizations have also flourished.

The EPRDF's federal policy has altered inter-ethnic relations in most parts of the country. The Guji and their neighbouring groups are some of the people who have been affected by this policy. In the Wondo Genet area, as discussed earlier, hostility and raids characterized the relationship between the Guji and Arsi Oromo. In these conflicts the Guji and Sidama allied against the Arsi, and the Arsi did not differentiate between the two in their fighting.

In 1992, there was a serious conflict between the Arsi and the Sidama. The major cause of the conflict was the question of where Wondo Genet should belong in the new division of the federal regions. Previously, the majority of Wondo Genet was under Sidama province, while a small part was allocated to Shashamene province of Shewa. In the new division, the Sidama and their new allies in the area (i.e. Kambata, Hadiya and Wolaita) wanted the whole area to be put under the Sidama Zone of the Southern Nations, Nationalities, and Peoples' Region (SNNPR), while the Oromo (both the Arsi and the Guji) wanted a large part of Wondo Genet (up to Bussa) to be allocated to the Oromiyaa Regional State. The government allocated most of the area to SNNPR, which triggered conflict between the groups. Consequently, the Guji signed a petition and submitted applications to the Prime Minister's Office asking for the land to be allocated to Oromiyaa. At the time of writing, no responses had been received, however.[3]

Using rights enshrined in the new 1994 Constitution, both regions changed the languages used in formal administrative settings and in elementary schools. The Oromo and Sidama languages (both written in Latin script) became the working languages of the respective regions. This action strengthened the opinion that only 'the people of the region', whose language becomes the working language, would control political power. The introduction of local languages in elementary schools made the issue more heated. The decision as to where an area belonged determined whether an Oromo child would be taught in the Sidama language or vice versa. The

---

3. Because of the continued tensions and wars in the area, the Ethiopian Parliament has recently decided to address the problem through referendum, which is now scheduled for 9 November 2008.

resentment of both groups who had no choice but to use the other's language made the contestation of the regional demarcation of Wondo Genet more vigorous.

One of the major new developments in inter-ethnic relations after 1991 was the change of alliances among the groups and the formation of new coalitions between some of the formerly conflicting groups. The Guji, who had been fighting the Arsi for years alongside the Sidama, changed their position and allied with the Arsi. The Kembata and Hadiya, who had previously fought with the Sidama, allied with the Sidama against the Oromo.

There are many reasons for changes in alliances. The influence of ethnic-based political organizations and the related revival of ethnic identities and ethnic nationalism were responsible for the new alliance between the previously conflicting Arsi and Guji-Oromo. The Oromo Liberation Front (OLF) and later the Oromo People's Democratic Organization (OPDO) cadres worked among the Guji to incite Oromo ethnic nationalism and to strengthen their Oromo 'ethnic feeling'. This played a crucial role in increasing the feeling of ethnic nationalism of the Guji in the area. The active political work and the revival of Oromo ethnic nationalism finally led to a reconciliation ceremony between the Arsi and the Guji, a ceremony in which notable elders from both sides participated. In the ceremony, according to Zerihun (1999), it was stressed that as both groups are fellow Oromos, belonging to one group and speaking the same language, the conflict between them so far had been an act of fratricide. From that day onwards both groups pledged to treat each other as brothers.

Undoubtedly, the Guji and Arsi also felt that they could gain political and economic advantages by working together. The introduction of the local languages in offices and schools continues to give them cause to unite. Guji children growing up in areas now defined as Sidama still use the Sidama language as their medium of education, perpetuating feelings of resentment.

In spite of the formation of this new ethnic alliance, some individuals have maintained their old inter-ethnic alliances, even though their older friends are now seen by their wider ethnic group as enemies. Some of the Guji and Arsi living deep in the Sidama-dominated areas and/or who are affiliated with the Sidama avoided being drawn into the conflict. Place of settlement, social bonds (through marriage and blood) and fear of possible retaliation hindered them from participating in the fighting.

In a similar way, the Guji who lived harmoniously with the Gedeo came to blows following the post-1991 political changes. In 1998, conflict broke out between the Guji and the Gedeo, which had at its root the gradual but steady expansion of the Gedeo into the Guji territory. The expansion was due to increase in their population and due to ecological factors. In previous decades, the Gedeo had settled as tenants on Guji land. Following the 1975 land reform, the Derg administration converted these tenant rights to usufruct rights, disenfranchising the Guji in what became about forty PAs. Later, with the fall of the Derg government and the establishment of a federal structure, conflict arose over the question of where these Peasant Associations should belong in the new division of the Federal Regions.

The Guji believe that the land belonged to their ancestors and consider themselves to be legitimate owners. They could not tolerate the idea of allowing latecomers (intruders) to become newly recognized legitimate owners of their land. Thus, the conflict culminated in a war. In the war, as mentioned earlier, many lives were lost, property was destroyed and thousands were evicted from their homes. The tension, suspicion and mistrust that have developed between the two formerly friendly groups has meant that the problem has become protracted and has not been solved up to the present today.

In the meantime, under the new federal policy and structure, the three Oromo groups (the Guji, Borana and Arsi), who used to consider each other as *Siddi Saddin* (the three enemies), were put together under the Oromiyaa Regional State. They became reconciled with each other as Oromo 'brothers' and formed a new alliance among themselves.[4] They fought together against their new 'enemies' such as the Gedeo, Garre and other Somali groups. In 2001, in a conflict between the Garre and the Borana (see Schlee, this volume), the Guji and Arsi fought on the Borana side against the Garre and their allies. The inter-ethnic relations between the Guji and their neighbours have therefore shown dynamic changes with the macro-level political changes. Changes in the macro-level political conditions have brought about changes in micro-level inter-ethnic relations and alliances.

Conclusion

This chapter has described the changes in inter-ethnic relations and alliances between the Guji-Oromo and their neighbouring groups in southern Ethiopia. Inter-ethnic relations between the Guji-Oromo and their neighbouring groups ranged from conflicts, wars and raids to cooperation and symbiotic relations.

In the south, the relationship of the Guji with the Borana and Arsi Oromo used to be hostile. They used to consider each other as *Siddi Saddin* (the three enemies) and fought each other for economic, social and political reasons. Guji relations with other groups like the Sidama, Burji, etc. in the south was not harmonious, although it was much better than those with their Oromo neighbours. The exception to this was that the Guji used to have cordial and symbiotic relations with the Gedeo.

In the north (i.e. in the Wondo Genet area), Guji relations with the Sidama were harmonious and they used to unite in alliance against their common enemy, the Arsi. However, owing to changes in government policy and growing ethnic self-consciousness, inter-ethnic relations and alliances between the Guji and their neighbouring groups have been radically transformed.

The new Guji and Arsi Oromo alliance in the Wondo Genet area and the Guji, Borana and Arsi Oromo alliance in the south are mainly the result of the new government policy, which acknowledges ethnic identity and ethnic rights. National political systems and state policies can significantly influence inter-ethnic relations and alliances at local level. Modern states control and allocate resources to different users and, as a result, directly or indirectly control them. Alfred Darnell and Sunita Parikh (1988: 270) comment on the role of the state as follows: 'Perhaps the most

---

4.   However, in later years conflicts and wars have re-emerged between the Guji and the Borana Oromo.

influential external force on ethnic groups and ethnicity in modern times is the state. As a major actor in social processes, the state is in a position to define the terms and contents of interaction within ethnic groups, between ethnic groups, and between an ethnic group and state.' In our case it was the adoption of the new federal system of administration that altered inter-ethnic relations and consequently inter-ethnic alliances in the study area.

However, ethnic identities are not always consistent and singular. Groups or individuals could develop different levels of ethnic identities, which are used in different places and contexts. To explain the hierarchy of ethnic identities Jenkins (1996: 816) writes, 'In as much as it is situational, ethnic identity is also likely to be segmentary and hierarchical. Although two groups may be differentiated from each other as A and B, in a different context, they may combine as C in contrast to D.' Likewise, the Guji, Borana and Arsi on the one hand regarded themselves as distinct from each other and, consequently, developed distinct lower-level ethnic identities as Guji, Borana and Arsi. On the other hand they belonged to the larger Oromo ethnic group, spoke the same language and developed a common identity as Oromo. In the latter context, they disregarded the boundary between them and embraced a common Oromo ethnic identity. In reality, they used both identities (i.e. the lower levels as Guji, Borana and Arsi and the higher one as Oromo) situationally. Thus identity can be wider or narrower depending on conditions.

In the recent conflicts against the Sidama, Gedeo and Garre, the high-level identity (Oromo) has been evoked to unite the Guji, Borana and Arsi. However, besides the common ethnic identity, the expectation of real or imagined social, political and economic benefits has had a crucial role in achieving this alliance. Moreover, while ethnicity is a form of social organization, it is not a hard and fast means of mobilizing each and every member of a given group. An individual's way of life, area of residence, economic occupation, degree of involvement in other non-ethnic networks and kinship ties all influence his or her reaction to ethnically motivated actions and responses. In the 1990s inter-ethnic conflicts, some individuals remained passive in the conflicts or sided against their own ethnic group members. In general, the data from the study area indicate that inter-ethnic relations and changes in ethnic alliances are determined by local socio-economic and cultural factors as well as by national policies.

Part IV
Pastoralists in the
Kenya–Ethiopia Borderlands

Chapter 11

Changing Alliances among the Boran, Garre and Gabra in Northern Kenya and Southern Ethiopia

Günther Schlee

After the elections of 27 December 2007, Kenya was beset by ethnic violence. The rival presidential candidates had ethnic constituencies, and, as violence along these ethnic boundaries erupted, many scores that had nothing to do with the presidential elections were settled. Most of them were about land issues, others about business opportunities and employment. There have been large-scale expulsions of people, reminiscent of 'ethnic cleansing', a term coined in association with Bosnia in the early 1990s. Even if the present attempt at power-sharing results in lowered levels of violence, Kenya has taken further steps towards being a country divided along the lines of territorialized ethnicity.

This recent conflict has emerged in the high-potential, densely populated areas of central and western Kenya. In contrast, the present chapter refers to the arid north of Kenya, an area where (before the 2007 elections, which were hotly contested) high-ranking politicians from the more developed southern parts of the country rarely bothered to visit to campaign for votes. The vast areas are thinly populated, and the few votes one could get there were not considered worth the effort of travelling long distances along bumpy dirt roads in the heat and dust. The material presented here, moreover, refers to the time around 2000, seven years before the recent escalation of ethnic violence, a time when Kenya was still considered an island of relative peace and a model for development. The chapter describes how, despite this relative calm, in a 'normal' situation in a remote area of Kenya, said to be 'peaceful' because violence remained at levels so low that it did not make it into the international media, there has been a progressive ethnicization of politics.

The Boran-Oromo were a dominant force in what is now southern Ethiopia and northern Kenya from some time in the sixteenth century to the threshold of the twentieth century. Their decline began when Menelik's armies arrived from the north and the British from the south. Previously, under the Boran-Oromo ritual umbrella, and within their system of military alliances, known as *Worr Libin* (the 'People of

---

1. The same as Liban discussed by Kellner, this volume.

Libin', after an area in southern Ethiopia[1]), they combined with many Somali-like and Somali groups, among whom the Oromo language spread. The history of this integration was conflictual: some groups submitted to Boran hegemony sooner than others; while others remained hostile outsiders to the *Worr Libin*. These differences in alliances led to splits across clans, as well as to new inter-ethnic relationships. The formations that emerged from these divisions and alliances have had quite different political fates and have undergone quite different linguistic and cultural developments. These complex inter-ethnic clan identities are the subject of a book (Schlee 1994a [1989]), but the main forms of political identification that have resulted from this history can be summarized as:

1. A Boran/Oromo/ *Worr Libin* identification.
2. A Somali identification that goes back to earlier forms of being Somali. In the literature, groups that claim these distant relations to Somali are referred to as Somaloids or Proto-Rendille-Somali (PRS). Before the sixteenth century, many groups of northern Kenya who now speak the Boran dialect of Oromo spoke a Somali-like language.
3. An Islamic identification, which in local discourse is associated with being Somali, or being like Somali (in this context the contemporary Somali neighbours), although by now a large number of the Oromo are Muslims as well.
4. An identification with the age-grade system of the Boran; here, in response to the strong discourse of Islamic Somali nationalism, the age-grade system of the Boran is seen as a religious institution. (Other Boran reject this identification and would rather compete with their neighbours for religious prestige in Islamic terms: 'If a purer Islam is something they have got and we have not, let us take it from them!')
5. An identification through inter-ethnic clan relationships that tie the descendants of the Proto-Rendille-Somali early occupants to each other. Only a few lineages among their present derivates claim Boran decent. They share with the Boran only a second set of ritual clan-to-clan relationships called *tiriso*, which was known to be an adoptive kind of relationship. *Tiriso* fell into oblivion after the decline of Boran hegemony over the twentieth century. These inter-ethnic clan relationships and other elements of shared history provide the basis of a *Worr Dasse* identification. *Worr Dasse* are 'the people of the mats', i.e. the Somali and Somaloid camel nomads, who live in mat-covered tents. These are distinguished from the Boran cattle-keepers, who live in grass-covered huts and are therefore called *Worr Buyye*, 'people of the grass'.

These five modes of political identification can be drawn on at different times by the same people. A modern Gabra, for example, most of whose ancestors probably spoke a Somali-like language before the sixteenth century, can either stress the old 'Somali' links of the peoples deriving from the Proto-Rendille-Somali (mode 2), the shared historical experience with other groups derived from the PRS (mode 5), the fact that he speaks Boran and can point to a close political affiliation to the Boran in the past (mode 1), or that he is a Muslim (as many Gabra of the northern Miigo cluster are) and under that aspect he would very much be presenting himself as like his more powerful Somali neighbours (mode 3).

Further factors of importance in the regional identity games are the national identities as Ethiopians or Kenyans. (Leaving aside the Somali irredentism of the 1960s and 1970s today there is no state by the name of Somalia which any inhabitants of territories outside it would like to join. Being Somali, defined in ways aside from the nation state, however, continues to play an important role at sub-national and transnational levels.) Of course, identification with the nation states that took possession of the areas has changed over time with the changing policies of these states towards the local populations. These relationships with the state have been marked by periods of war or relative peace, and the states themselves have been through various different phases: from imperial to socialist to ethnocratic in the Ethiopian case; from colonial to post-colonial to ethnocratic in the Kenya case.

Having written extensively on the colonial and post-colonial settings in connection with nomad/state relations (Schlee 1998a, 1999), I want to concentrate here on more recent events, from the 1990s to present, which, as far as state policies are concerned, are influenced by new roles accorded to 'ethnicity' in both Ethiopia and Kenya. Most of the data I refer to, from interview texts and press cuttings, have been collected by Abdullahi Shongolo and will be part of a book-length study (Schlee and Shongolo, in preparation). I can therefore take the liberty here of omitting original texts and more detailed documentation and concentrating on drawing an outline of the shifting alliances. Among the *Worr Dasse*, the groups involved in recent conflicts and among whom there have been shifting alliances since the 1990s include the Gabra, Garre, Ajuran and Degodia.[2] Here I am going to concentrate on the Gabra and Garre and their relationship to the Boran. In another publication (Schlee 2007) I focus on the Degodia and Ajuran. As all the conflicts in the area are interrelated, there will be occasional references to these others. As the Ajuran have in recent years become allies of the Boran, while the Degodia have become their enemies, a brief word about them is necessary even at the start of our discussion.

Degodia and Ajuran are at first sight culturally indistinguishable. There are Boran-speaking Ajuran, but the remainder of the Ajuran speak the same form of Somali as the Degodia. When I visited Somali hamlets in the eastern part of Marsabit District and in Wajir District in 1977–80, I found Degodia and Ajuran living in intermingled settlements and there was no way of telling who was who without asking. The relationship between Degodia and Ajuran then deteriorated, culminating in massive raids and rape and murder. In 1984, the Ajuran managed to instrumentalize government forces, and hundreds of Degodia were kept as captives on Wagalla airstrip

---

2. Degodia are 'people of the mats' in the wider sense of camel nomads. They are relatively recent arrivals and have not undergone the same history of Oromoization as other *Worr Dasse*.
3. The Darood originate from further north and east and entered the Kenyan scene only in the early twentieth century.

near Wajir, where they perished. Darood[3] Somali from Garissa, both herders and people in government office, also added to the pressure against the Degodia.

The year 1984 was very dry and pressure on pastoralist resources was high. Darood Aulihan Somali from Garissa District pressed into Isiolo District and started to squeeze the Waso Boran (i.e. the Boran who live along the River Ewaso Ngiro east of Isiolo in Kenya) from their pasture-lands. Many Boran herds were raided. At the same time, Degodia had moved into Isiolo District, partly because of the drought and partly because of the Ajuran threat. There, they managed to establish a better relationship with the local Boran, and they helped them to retaliate against the Darood Aulihan. That they helped their 'traditional enemies', the Boran, against their fellow Somali is not surprising if one considers what they had suffered at the hands of the latter. The consequence was significant, however: the Waso Boran informed the Boran of Marsabit and Moyale that henceforth the Degodia should be regarded as allies.

Some Degodia, who had already picked up bits of the Boran language in the Waso area, when asked who they were, started to identify themselves as Boran:

'*Att gosi maan?*' '*Booran.*' – 'Which tribe are you?' 'Boran.'
'*Balbal tam?*' '*Degodia.*' – 'Which section [gate]?' 'Degodia.'

Degodia thus, at least in some situations, became equivalent to a section of their own 'traditional enemies'. In one of the ad hoc genealogies that sprout in such situations, the Degodia were not depicted as descending from the Boran but as a branch collateral to them: both 'Degodia' and 'Boran' were treated as eponymous ancestors and were said to be the sons of one ͨAli. The Boran tolerated this realignment and took it as a sign of goodwill, although it was contradicted by other genealogies, such as the one attributing to 'Boran' and 'Waat' (the putative ancestor of the former hunter-gatherers) a shared father 'Horro'. The Gabra, who had a long-standing *tiriso* relationship with the Boran without ever having claimed common ancestry with them, were amused by this new genealogical fancy.

Nothing derives from its roots. A plant grows from the surface, where its seed has come to lie, both ways, up and down, at the same time. When the root metaphor is used in the context of human history, it is mostly used wrongly. Peoples and states are said to stem from their roots. In reality, they drive their roots from the present into the past, from the surface of time into its depth, just like a plant drives its roots downwards into the soil. The Degodia were trying to grow roots in a new soil, as Somali groups and other lineage-based societies have always done, by linking their genealogies to some real or imagined ancestral figures of the local society. Had this genealogical fiction persisted long enough, people might have started to believe it.

The Degodia gave some substance to their new relationship with the Boran by guaranteeing them safe passage from Moyale to Isiolo as the crow flies. The road diverts from this line towards the west, and had become obstructed by so many police barriers, where lorries, with and without insufficient documentation or other faults, had to pay bribes, that road transport had become very costly. Boran traders, and Burji traders, who were not distinguished from them, took advantage of the

Plate 11.1 A Garre hamlet on the move in Mandera District, Kenya, 1990 (photo: G. Schlee)

Degodia guarantee of security and trekked herds by foot to the central Kenyan markets by the straight route through the bush. This route came to be known as the Bosnia route. In this period, the words 'ethnic cleansing' and 'Bosnia' were frequently heard together on the transistor radios that link these mobile populations with the outside world. They used this new name to infer that, as the new route had been cleared of Ajuran, it was believed to conform to Bosnian standards.

The Degodia did not distinguish between Boran and Burji traders, but the Boran traders did. The Boran tended to despise the Burji, who were originally mountain farmers and poor in cattle. They did not find their new role as highly successful competitors in business endearing (see Amborn and Kellner, this volume). In order to prevent what they saw as evil associated with the Burji accompanying them on their route, the Boran sacrificed a billy goat (*korbes*) at a place called Mad'eera Kaayo. They did not, of course, publicize this to the Burji.

In the same measure that the relationship between the Degodia and Boran improved, the relationship between the Garre and Boran declined. In a way, the Degodia came to occupy the niche formerly occupied by the Garre, as camel pastoralists in the Boran orbit. In the Negelle area of Ethiopia, raids occurred between Garre and Boran, and also the Degodia took advantage of this situation and raided the Garre there. These were not the same Degodia as those who had come from Isiolo District to Moyale, but people of the same Somali clan who had been living in Ethiopia for some time. Cattle taken as loot in these raids were taken to Kenyan markets. The Kenyan Garre became convinced that any cattle taken by Moyale traders to Nairobi by lorry rightfully belonged to them. Influential Garre succeeded in influencing the Veterinary Department to close the road to cattle transport under the pretext of the outbreak of some disease somewhere. These transport restrictions 'for veterinary reasons' have a long history of being misused for political and economic purposes (Schlee 1998a). However, the closure of the road also furthered the diversion of trade onto the 'Bosnia route' and strengthened the interdependence of the Boran and Degodia, to the detriment of the Garre.

The year 2000 brought perpetual new variations on old themes. Violence escalated mainly around Moyale and Isiolo and the areas adjacent to the east of Mandera, Wajir and Garissa Districts, and double-digit numbers of casualties in single confrontations became a regular feature of the news. Apart from the usual level of livestock rustling, Marsabit District was relatively calm. Marsabit town had had its share of urban violence, with clashes about agricultural land in the vicinity of the town in 1998 and killing between Boran, Gabra, Rendille and Burji.

In 2000, the main line of confrontation was between Boran on the one hand, and 'Somali' of various categories and degrees of 'Somaliness', if that term is permitted, on the other. In Moyale some *Worr Dasse*, the 'people of the mats' (i.e. the camel pastoralists or former camel herders), who once had formed the *Worr Libin* alliance with the Boran as senior partners, now sided with the Boran again. The Ajuran, whose gradual defection from the Boran in the 1920s and 1930s has been described elsewhere (Schlee 1994a [1989]: 46f.) now resumed their position of the 1910s again, appealing to the memory of their old political affiliation to the Boran. The Garre did not. They remained aloof and became viewed as the embodiment of the threat of Somali expansionism in the Moyale context. In fact, it was violent competition over grazing resources between the Garre and the Ajuran that drove the Ajuran back into the arms of the Boran.

The Ajuran had found themselves opposed by other Somali, namely the Garre and Degodia. Competition had been aggravated by ecological conditions: the drought of 1997 was followed by the 'El Niño' floods of 1998, which decimated the weakened livestock. Since that deluge the drought had resumed. Vast areas of East and North-East Africa saw no proper rainy season in 1999 or 2000. With a diminished pastoral resource base, contested by Garre and Degodia competitors, the Ajuran became nostalgic about the once-privileged position they had enjoyed at the side of the Boran.

In addition to pasture and water, conflict over resources associated with urban life and the national state played a role. In Moyale, in addition to recent pastoral and agro-pastoral Garre immigrants, there was also an old Garre business community, which had been there from the very beginning of the town.

In the first phase, the violence took place mainly on the open range, in the grazing areas in Wajir District and in the eastern parts of Moyale District. In March 2000, conflict followed a disagreement between Garre and Ajuran about the appointment of a chief for the El Danaba location and a subchief for Irrees T'eenno. The area had hitherto been dominated by Ajuran. Now the Garre claimed these offices. A Garre was killed near Gurar on the 16 March. Then a mine exploded at D'okisu in Ethiopia, killing twelve Garre. The Ajuran were suspected of planting it, and, on 22 March one Ajuran was killed by a Garre near Gurar. Then two Ajuran hamlets were burnt. On 27 March, four Garre were killed at Qofoole in Wajir District. Then the killings spilled over into Moyale District. A lorry, a Canter, belonging to an Ajuran businessman was attacked on the Moyale-Dabel road. There was an exchange of fire between the escort of the lorry and the bandits, who were believed to be Garre. The same evening two Garre were killed at Nana and their genitals were removed. One of the bodies, that of an elderly man, was said to have an Ajuran property mark carved into its thigh. Garre now started to flee into Ethiopia, while Ajuran sought safety to the south and moved in the direction of Isiolo. Killings continued on this scale throughout early April. In mid-April there were major clashes in Moyale town itself.

On 4 April 2000, there was a demonstration of Garre in Moyale town following the arrest of two elders who had been accused of fuelling clashes between Garre and Ajuran in the northern part of Wajir District. Demonstrations and riots continued in the following days. In the course of these demonstrations one Kenyan flag was burnt on 5 April, an incident that gave rise to rumours that the flag of Region 5, the Somali region of Ethiopia, was raised in its stead. Under the pressure of these demonstrations a cash bail of Ksh 30,000 each was accepted for the two elders on the initiative of Haji Ali Barricha, a Mandera politician from the oppositional Ford Kenya party. Haji Ali Barricha was a Garre, who had come with a lawyer and a group of human rights activists. Then, on 7 April, a young Garre was shot by a policeman as a riot developed between the Garre demonstrators and Gabra and Boran youths. Ali Barricha was informed of the event and came again, this time by aeroplane with newsmen, human rights activists and a pathologist to perform a post-mortem. The group was prevented, however, from proceeding from the airstrip to town by Boran and Gabra, in collusion with the District Commissioner. The next day, stone-throwing between mobs from the two sides continued and shops were looted. On 10 April, Dr Guracha Galgalo, the Member of Parliament for Moyale and an Assistant Minister for Health, a Boran, arrived on the scene. In a meeting with Gabra and Boran elders the following day, he stressed the anteriority of the Boran and Gabra in the area. The Sakuye, another group which, like the Gabra, is of mainly Proto-Rendille-Somali origin, was simply subsumed into the category 'Boran'. The Garre were accused of expansionism. Dr Guracha's former rival candidate for parliament, a Garre, was quoted as having said: 'Our grandfathers took the Acacia of Warraab [in Somalia], our fathers took the Acacia of Hariyya [in Ethiopia, near Lae, a short distance north-east of Moyale], and it is us who will take the Acacia of Obbu [near Sololo, in Kenya, beyond Moyale to the west].' Dr Guracha also expressed a suspicion that the local representatives of the Kenyan Government did not care enough about their flag being burnt. He also claimed that the Boran business community in Moyale would suffer more from violent conflict than the Garre businessmen, because the latter could simply move to Mandera or Wajir.

On the following day, 12 April 2000, an aeroplane was expected from Nairobi bringing journalists and the lawyer hired by the Garre. A delegation of local politicians and businessmen wanted to stop these people at the airstrip, to avoid negative press coverage for Moyale. Among them was Dr Guracha. The plane also brought the newly appointed District Commissioner (DC), Clement Nzoimo

Kiteme, and Ali Barricha, the Garre Ford Kenya official from Mandera. The incoming DC was introduced by Barricha to the area MP and to the other leaders on the airstrip. That is how they first heard about the fact that now they had a new DC. They were infuriated by the fact that the new DC was introduced to them by an opposition politician and concluded that he must be pro-Garre and anti-KANU.[4] The controversy became so heated that it led to blows between the County Council Chairman, Mr Golicha Galgallo and Mr Barricha. Mr Barricha was also alleged to have handed over a newspaper to someone in which a map labelled 'Garriland' was found, which led shortly after to further bitter controversies, for, as we shall see later, totally absurd reasons.

Following the news of the DC's transfer, the talk of the town was that the Garre community had overpowered the so-called indigenous community, by being able to bring in a new DC of their own choice, and allegedly a Ford Kenya DC. The outgoing DC, Mr Kipkebut, had been accused by the Garre community of siding with the Gabra/Boran.

During the week from 7 April to 13 April, the Kenyan/Ethiopan border was closed at Moyale by the district administrations on both sides, after two days of joint military operations. The closure was aimed at preventing Ethiopian Garre from assisting their fellow tribesmen living directly adjacent to the border on the Kenyan side. Because of the economic interdependence of this border area, the economic effects of this measure were immediately felt on the Kenyan side. The price of a tin of maize went up from Ksh[5] 13 to 20, that of a kilogram of sugar from 40 to 60, a cup of milk from 8 to 20 and a bundle of *miraa* (a mild vegetative narcotic similar to *qat* or *chat* in Ethiopia and Somalia) from 20 to 35.

A curfew was also proclaimed in the Kenyan town of Moyale from 1800 to 0600 hours, a directive that reminded the older generation of the 1960s and 1970s, when Moyale was under curfew for about eight years. The children born in that period are still known as *ijolle kafio*, the 'children of the curfew', and said to be particularly numerous 'because husbands had nowhere else to go at night'.

A Garre perspective on the new front lines of these ethnic clashes is provided by an interview Abdullahi Shongolo carried out with Councillor Adan Sheikh, the County Council member for Nana location, on 14 April.[6] When asked about the cause of the quarrel between Garre and Boran, the councillor talked ominously about a spider-web between the two sides that has brought them up against each other, and went on to say: 'You know, a divorced woman who did wrong to her husband and left will not do any good to the next one. The Boran normally do not understand things [quickly], we shall come to see that this divorced woman will cause damage to them.'

---

4. Kenya African National Union, then the ruling party.
5. Kenya shilling.
6. A little over five months after the interview, on the night of 24 September 2000, Adan Sheikh was murdered by unknown gunmen outside his house as he came home from the mosque (East African Standard, 26 September 2000, p. 4).

Both the 'spider-web' and the 'divorced woman' are a reference to the Gabra (Miigo). The 'spider-web' is meant to describe their weakness and futility, while the 'divorced woman' refers to the circumstances that, whereas before Gabra sided with the closely related Garre, from whom they might have separated over some quarrel, now they side with the Boran. He continued:

You know, the Gabra [using the Amharic form *gabarti*, which stands for dependents of low status] have the habits of poor people. Have you ever seen the children of poor people who have grown up? If you give everything they want to the children of the poor, they will still treat you badly. In the end you will die because of them. Now, these people are like that. If they were not between us and the Boran, we would not quarrel with the Boran.

On the evening of 12 April, there was a report on BBC radio. It is claimed that this report said that 90 per cent of the population of Moyale District were Garre. Rashid ʿAbdi, a private telephone operator, was suspected of having given the BBC an interview. He was called to Security and admitted to having provided this information.

In the resulting atmosphere of increased sensitivity, the matter of the map that had been found between the pages of Barricha's newspaper came up. Photocopies of it were sold in several shops for Ksh 50 a piece. It was a completely innocuous map from a book by Shun Sato (Sato 1996: 276), which is meant to give a rough orientation of where the Garre live. Many other ethnic groups are mentioned as well in their approximate locations. The legend clearly states that the uninterrupted bold lines on this map are the main roads. Nevertheless, there were heated discussions in which the roads leading from Dila in Ethiopia to Dolo (where the borders of Kenya, Somalia and Ethiopia meet), to Wajir and up again via Moyale and Mega to Dila were pointed out as the boundaries of 'Garreland'. It was said that the Garre claimed the whole area comprised by this triangle as theirs. The literate people who forwarded this interpretation must have known that it was a misreading.[7]

The alleged territorial claims by the Garre led to heated debates. In a meeting attended by selected elders of Garre, Gabra and Boran (who together were known as the District Security Committee or DSC team) and delegates from Nairobi led by the MP Dr Guracha, Councillor Abdirahman, a Boran, gave a speech in which he described this map as proof that the Garre illegitimately claimed a large area of land. A former chief was called upon to explain the matter to the audience:

If we look at this map, inside Kenya the name Boran does not even appear. The name Boran is put between Yaballo [Yabelo] and Mega. The Somali [Region 5?] boundary inside Ethiopia goes from Waachille via Moyale to

---

7. Georg Haneke, a student of Schlee, who was in Moyale in June, still found people debating this map. He pointed out this misreading to them.

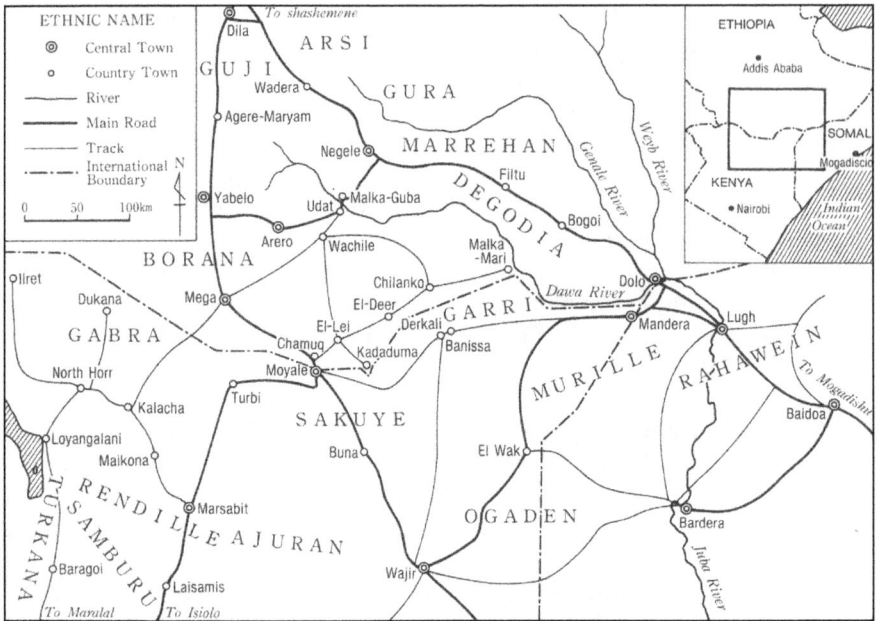

Map 11.1 Garriland and the tripartite borders of Ethiopia, Kenya and Somalia (from Sato 1996: 276).

Lag Warabeesa [near Turbi in Kenya, pointing to other roads]. From here it is drawn on to Buna, from there to Dolo. From Dolo it is drawn to Waachille. That is the map.

The reader can easily convince him- or herself that there are no internal boundaries of any state at all on the map, because we reproduce it here (Map 11.1). There is not even any mention of the Somali area of Ethiopia.

But pressure on the Garre delegates during this meeting became so strong that they had to apologize for everything that was alleged against them, including the plainly absurd allegations involving this map, and had to pay Ksh 5,500 as an appeasement (*hoola*, literally: a young female sheep) to the Boran and Gabra. The Garre elders were also under time pressure to bury their dead. Being aware of this situation, the Gabra and the Boran used this opportunity to corner them in the debate and to pressurize them to accept guilt for having caused violence in Moyale District.

The pressure on all actors of the Moyale scene and the heat of the controversy were increased by occurrences taking place simultaneously. There were clashes between Garre and Ajuran in Wajir District and between Garre and Degodia at Malka Wiila near Dolo in Ethiopia, where thirty-seven Garre were reported dead. On 12 April, there was a demonstration in Wajir by Degodia, Ajuran and Ogadeen Somali, who demanded that the Garre should leave that district. There were further

---

8. On Garre as refugees, see also Getachew, K.N. (Getachew Kassa) (1996).

killings and Garre fled towards Moyale. On 17 April, seven lorries full of Garre refugees arrived at Moyale, which did not ease the situation.[8]

While the Ajuran found their way back into the *Worr Libin* alliance under the umbrella of the Boran, the Garre moved the opposite way. They identified themselves more clearly as 'Somali' than at any time before. Being Somali was equated with being Muslim, and their political rhetoric was that of *jihaad*. During a demonstration in Moyale, Garre youths and women sang:

> *Olkiin arra jihaad. Nami kufaar tokko ijjeese reer jannaada.*
> 'The war of today is *jihaad*. Whoever kills one infidel will belong to the people of heaven.'

The following text is from an audio cassette, produced in Ethiopia and mainly referring to Ethiopian events, which sold for Ksh 200 in many places in northern Kenya and Ethiopia in April/May/June 2000. It was also available in Nairobi. The cassettes were advertised by word of mouth and at a later stage distributed in the small kiosks that also sell *miraa* and cigarettes. The A side of the cassette contains Islamic instructions about the *ahaadith*. Only the B side, which we quote here, is political. As it was thought that the content might be offensive to Boran, at first this cassette only circulated among Garre. The political poetry shows stylistic similarities to the poetry of Jarso Waaqo Qoto, the Boran poet (for samples of his poetry, see Schlee and Shongolo 1996; Shongolo 1996). The poetry illustrates the point that adversaries may be, or often become, similar to each other:

Obboleeyan tiyya	My brothers,
diqqaa guddaan nu kaasa	young and old, let us rise
jahaada raasatti nu baasa,	let us go to the bush for *jihaad*
waan harka qabdaniii	with what you have in your hands
gaalki onne isa tarsaasa,	cut the hearts of the infidels to pieces
haalo aanan gaal keenna	the vengeance of the milk of our camels
maati isa marsaa hobbaasa,	round up his family and finish them
gaalki d'iig isa lolaasaa	make the blood of the infidels flow
allaatifi waraabessa nyaacisa,	make vultures and hyenas eat them
haga badda Magaadalle	up to the highland of Magaada
oriisumaan gul nu yaasa,	let us chase them in one go
chiniinsatti chiniinsa sirra bukhisa	pain is driven out by pain
hand'uraati duuda dubbisa,	the legal property makes the dumb speak
nu akkamiin teenne laalla	how can we just sit and watch
ka amma gaalkin aanan gaal keenna	the infidels lie down satiated with the milk of our camels
chiise ilmaanin guddisu,	and raise their children on it
gaalki haati badde galat Rabbi keenna.	thanks to the Lord, the mothers of the infidels are lost

arra Garrii la beekatte	today the Garre have become aware.
la gadi baafate	they have come out
bineeyin arra haalo teen baasitu	the wild ones who will carry out our vengeance today
ta d'iig kanke lolaasitu	who will make your blood flow
ta jagna ke kolaasitu	who will castrate your brave ones
ta Raaba gada atin abuudulle	who will make the *raaba gada* you believe in,
finchaan harree obaasitu	drink the urine of donkeys
ta qutum ke roraasitu	who will shake your settlements
ta haate lafa si baasitu	who will attack you and chase you out of the land
ta guulki Garrii ol aansitu.	those who make the victorious Garre rise to the top.
Si'i farsoofi bakhti nyaatu	You who drink alcohol and eat unclean flesh
ka boqorkiin ke waraabes waliin	whose leaders compete with hyenas
bakhti saamu	over carcasses
badiiti si tolce male	it is because you are going to perish
ati guulkii akkan hin taatu	you are not doing so for victory
wakhtaati d'io jira	the time is close by
ka geesigiin keen d'aadatu	when our skilled ones will proclaim
ka raabi ke guutu haadatu	that your *raaba* shall shave off their tresses
ka Abbaan Gadaallen ashaadatu.	and the *Abba Gada* will pronounce the *shahaada* [the two articles of the Islamic faith, i.e. convert to Islam].
Sabbo sadeen ijjeesi	Kill the three Sabbo
akk injiraani	like lice
Gona torbaan ha goolani	let the seven Gona be burnt
akka qoraani	like firewood
mirga bita uf laali	look out right and left
d'iig ke ulaan barbaadi	for a place for your blood
Gardallo martitti	Insolent foolish people
akka duul allaatii	like a swarm of vultures
waan guddo sii taate	did a lot for you in the past
duri Mangistuufi Amarti	in the time of Mengistu and the Amhara
fula warri keesa kae	in the place these people left
at teete barbaada	you want to establish yourselves
ibidaati d'umatte	you will perish in the fire

Plate 11.2 Trying to handle a loading camel: Garre, Mandera District, Kenya, 1990 (photo: G. Schlee)

imiimi rooba taata	you will become like the tiny insects which come out after the rain
Gardallo daalle	Insolent foolish people
akhli malaasa	with brains full of pus
me uf laallad'i	mind yourselves
akhli te doomaa	your brains are hornless
eesatti d'eebu baata	where will you quench your thirst
guyya jaba bona	at the time of a severe dry season
tulla saglaan	the nine great wells
amma d'iiratti tooa	now [real] men draw water from them
Dagaalkin waan jiran	The struggle has been there
saban Nabiyaashi keenna	since the era of our Prophets
yo dagaalkii doonelle	even if we die in this struggle
qubqabaadd'a qar reer jannaada	know that our souls are already
nafseen teen.	among the people of paradise.

216 Günther Schlee

Soomali gaal keen	Yellowish are our camels
Somaliin gos teen	Somali is our tribe
Hassan maqa keen	Hassan is our name
ha nuu taatu	let our repentance
tooban teen.	be accepted.

This text, which has the form of a sermon and the content of a harangue, equates the struggle of the Garre with a *jihaad*. A closer look, however, reveals not only that Muslim and Garre nationalism are mixed, but that many elements of the Cushitic 'killer complex' (see Schlee 1994b: 135) creep in. Islam explicitly forbids the mutilation of slain enemies, a practice that offended the Prophet at the battle of Uhud. The threat of castration, which is an allusion to the practice of taking genital trophies, is definitely un-Islamic.

There are other observations that contradict the simple formula Boran = infidels, Garre = Somali = Muslims. In the early 1990s Boran withdrew from mosques whenever their community was at war with Somali. The Garre, then as now, referred to the Boran as *kufaar*, 'unbelievers'. Then as now, this derogatory designation was applied, although many of the Boran were and are Muslims. In the early 1990s, however, the Boran mirrored the Garre view of them, by primarily identifying Islam with being Somali. Under attack by Somali groups, the non-Muslim Boran and the only superficially Islamized ones would distance themselves from Islam and criticize practising Muslims in their ranks as following Somali customs. In the recent clashes they did not do so. Islam was no longer regarded as a Somali attribute. Instead, even in this period, many Boran had converted to Islam as far inland as Dubuliq (in the central Boran region of Ethiopia). With the pan-Boran solidarity strengthened by the Garre threat, Muslim Boran scholars from the Waso area (Isiolo District) continued to preach and proselytize all over Boranland.

One such Muslim scholar, a Boran from Isiolo, Sheikh Abdullahi Golicha, was in the area preaching at the time of the clashes. The Garre community of Moyale, who declined even to pray behind a Gabra Imam in one of their mosques, invited him to preach to them. In his sermon, which was partly in Swahili, he reminded them that it was the month of Muharram, during which waging war is forbidden to Muslims.

Ninyi mnawaita Waborana na WaGabra ati wao ni kufaar. Badala ya kuleta hawa kwa dini ya Isilamu vile ume amuriwa na Mungu, mnawatorosha. Mwezi huu sio mwezi wa vita masheikh wenu wanajua hivyo. Hata katika aada ya Waborana, mwezi huu Waborana wenyewe hawaendi kupigana na maadui.

You call the Boran and the Gabra *kufaar*. Instead of bringing them to the religion of Islam as you have been ordered by God, you chase them away. This month is not a month for war, and your sheikhs know that. Also in the Boran customs, the Boran themselves do not go to fight enemies in this month.

On several occasions, Garre sheikhs joined him in preaching about Muslim brotherhood and about restorating peace to Moyale. Councillor Edin Adow of the Godoma location, one of the alleged Garre warlords 'wanted' by the Boran, attended evening prayer at Butiye, a predominantly Boran settlement, where Sheikh Abdullahi had come to preach. No one questioned him on this occasion.

After considering the Garre/Boran relationship it is worthwhile to look more specifically at the Garre/Gabra relationships in the light of the recent changes of alliance. Gabra Miigo (the northern Gabra to whom our sources refer in this context) have largely the same clans as the Garre. They share the PRS origin. Both groups are now Muslims, although the Gabra are said to be so 'only in town'. The ethnic split between them occurred when, centuries ago, the Garre managed to withdraw from Boran dominance for a while to the east, and the Gabra submitted to it with less delay. This is what, in modern parlance, makes the Garre more 'Somali' – although still in 1990 I spoke to many Garre who rejected that label – than the Gabra. The Gabra have also kept their peculiar form of the *gada* system (Schlee 1994a [1989]: 82f.) and, in connection with it, the ritual journey to their holy mountain Hees, on the top of which a camel has to be sacrificed, as distinguishing features.[9] As we have described earlier (Schlee and Shongolo 1995), in the early 1990s there was a split between the Boran and the *Worr Dasse*, 'the people of the mats', the camel-keepers who were the former allies of the Boran. At that time, the Gabra and Garre found themselves on the same side of the dividing line. Thus in 1992, during the conflict between the Boran and the Gabra/Garre, the Degodia had the opportunity to occupy the niche left by the two ethnic groups in relation to the Boran. The new Boran/Degodia alliance still worsened the rift between the Gabra and the Boran. As time went by, the Gabra realized that they had made a mistake by forming an alliance with the Garre, who did not allow them access to political representation, better pasture or more water points. The Gabra found themselves in the same predicament as before, during their decaying relationship with the Boran. 'They jumped from the frying pan into the fire or they jumped from the python's into the cobra's mouth,' as one elder explained. There was no way for the Gabra to return immediately into the Boran fold. An opportunity came when the relationship between the Boran and the Degodia deteriorated. To this, the Gabra might have contributed. Thus in 1998, the Degodia became a common enemy to both the Boran and the Gabra. Without considering their past differences, the Gabra at once renewed their alliance with the Boran to fight against the Degodia. As the Ajuran, who returned to the Boran fold as well, were at war with the Garre, the Gabra now found themselves in alliance with the enemies of their prior allies, the Garre. We now have Boran/Ajuran/Gabra against Garre and Degodia, the latter two also at odds with each other. The Gabra

---

9. On the holy mountains of the other branch of the Gabra, the Gabra Malbe, and the age-set promotion rituals associated with them, see Schlee (1990b, 1992) and Wood (this volume). These holy mountains have great significance as symbols of social identities. The rituals performed there and their timing further illustrate the interconnectedness of structurally quite dissimilar *gada* systems through the transfer of ritual services. The Garre, by the way, although they do not have a *gada* system of their own, are not outside this network (Schlee 1998b).

do not like the Ajuran occupying their old position, closest to the Boran. They see that the ecological niche they used to enjoy, on the outskirts of Boran grazing lands, is now occupied by Ajuran. It remains to be seen how stable this alliance will be.

We have so far looked at ethnic groups and district-level politics. We have seen that the rural, pastoral facets of the conflict in Moyale involved people from Wajir District and areas of that district. The international level so far did not play a major role, but soon the mutual raiding was also to affect Kenyan/Ethiopian relations. On 14 June 2000, the *Daily Nation*[10] (p. 9) reported that Ajuran took to Wajir streets the day before to 'protest a recent attack by militiamen from Ethiopia. Waving placards and chanting slogans, the Ajuran accused the Garre clan of masterminding the raid in which two people were killed and more than 10,000 head of cattle stolen last week in Wajir North … [T]he Government has lodged a formal complaint with the Ethiopian authorities over the attack.'

Under the headline 'Mission to Ethiopia Unsuccessful' the *Daily Nation* reported on 24 June 2000 (p. 24) that Kenyan leaders had only been offered the return of 160 animals which Ethiopian forces had confiscated from the raiders. As a result, 'the Kenyan delegation, which included the Wajir North MP, Dr Abdullahi Ibrahim Ali, refused to accept the animals, and insisted that the entire herd be returned'.

In fact, the number of losses given by the Ajuran might have been exaggerated. There were also rumours that, because some of the raiders wore uniforms, they might have been Ethiopian soldiers on an unauthorized mission, an off-duty income-generating activity, so to say. The Ajuran, however, in order to mobilize official Kenyan support, emphasized this Ethiopian involvement. They attempted to transform an affair between Kenyan and Ethiopian herdsman into an international affair. By shifting the issue to the international level they hoped to harness the power of the Kenyan state for their purpose. In the *Daily Nation* of 24 June 2000 (p. 7), we read: 'We earnestly appeal for President Moi's personal intervention on this matter so the Ajuran community can repossess their livestock. Unless this happens, our government will lose … legitimacy … It is the government's primary responsibility to protect the lives of its citizens and their property.' Also the Provincial Commissioner of the North-Eastern Province held Ethiopia responsible for this incident (*East African Standard* 28 June 2000, p. 9).

The Ethiopian embassy in Kenya sharply denied that Ethiopia had anything to do with the affair (*Daily Nation* 1 July 2000, p. 7): 'Ethiopia fully respects Kenya's sovereignty and strongly values the shared bond of friendship. The claims that Ethiopian soldiers have been making incursions into Kenya and killing innocent civilians are outrageous and not in keeping with the friendly relations between the two countries.'

As if education in northern Kenya were not plagued with enough problems in normal times, the insecurity led to the closing of schools (*Daily Nation* 7 July 2000, p. 19).

---

10. One of the main Kenyan national newspapers.
11. Another major Kenyan national newspaper.

Some of the pressure that local leaders had been putting on the government to take responsibility for the security situation in the area rebounded on them, however, as government officials started to blame the local MPs for provoking the clashes (*East African Standard*,[11] 9 July 2000, p. 28). In their response the MPs accused the government authorities of being inefficient. The evidence they cited was the government's inaction over the OLF. For a long time the government had continued to deny the presence of the Oromo Liberation Front in the area, despite reports by the locals about them.

The Provincial Commissioner of North-Eastern Province, which comprises Mandera, Wajir and Garissa Districts, announced that a leaders' meeting would be convened to discuss these 'clan clashes which were politically instigated and fanned by politicians' (*East African Standard* 7 July 2000, p. 28). This leaders' meeting, to discuss reconciliation of Ajuran and Garre, was first planned to take place at Wajir, but the Garre delegates refused to go there, fearing that their lives would not be safe there. Then it was to take place at Mandera, but the Ajuran found this unacceptable for similar reasons. The next proposed location was Garissa, the provincial capital, where neither the Garre nor the Ajuran felt secure. Finally Moyale, although it is situated in another province, Eastern Province, was agreed as a neutral ground. The meeting came as a surprise to Boran residents of the area, who were infuriated by it. They feared that the problems of the neighbouring province would be exported to them.

The meeting took place on 27 June 2000. The plan was that, after a closed session of the Provincial Security Committee, which consisted of government officers only, separate talks would be held with the Ajuran and then with the Garre, and then a joint session with representatives from both groups would be held. The plan received a severe blow, however, already in the Ajuran part of the programme. The Ajuran delegates walked out after declaring that they had no time for meetings announced at short notice: they had to go and bury their dead, whom the government had failed to protect, and to retrieve their looted livestock, which the government had been unable to do. There was, they said, no time for peace with the Garre now. They were confronted with accusations that they had received help from the OLF, which was alleged to have returned to the area, and the delegates responded provocatively in the affirmative: yes, even the OLF units from Sudan would rush to their help. (This is despite the fact that the Blue Nile area of Sudan, where the OLF once had bases, is about 1,000 km away from northern Kenya.)

On the national level, the insecurity in northern Kenya was used by the opposition to criticize the government. One can well imagine that regional administrators like PCs and DCs were in an uncomfortable situation between angry local elders and impatient superiors in Nairobi. The head of the largest opposition party made a press statement on 17 July (*East African Standard*, p. 8):

Government urged to negotiate on security

---

12. Later, in December 2002, Mwai Kibaki was elected President of Kenya; his contested re-election in 2007 caused widespread violence.

Leader of Official Opposition, Mwai Kibaki,[12] has asked the Government to immediately enter into talks with Ethiopia so as to end massacres and cattle rustling in North Eastern and Eastern provinces.

Kibaki wondered why the Government was quiet over the killings and loss of property that have rocked the area in the last three months. He said there was no need for President Moi to mediate between Eritrea and Ethiopia while his territory is suffering similar problems.

In fact the Ethiopian/Eritrean war to which the former Vice-President (and later President) Kibaki here alludes was not a separate conflict but was tied directly and indirectly to what was happening in Somalia and northern Kenya.

There were rumours in the area that the Garre had hired militiamen who had joined the RRA (Rahanweyn Resistance Army).[13] A similar perspective is reflected in an article in the *Daily Nation* 22 July 2000 (p. 24), in which it was described how 'raiders from Ethiopia and Somalia, backed by their [Garre] clansmen from Mandera District and armed with AK 47 assault rifles [Kalashnikov] and sub-machine guns' were involved in raids in Wajir District. Local MPs claimed that the 'breakdown of law and order in Somalia was the main cause of insecurity in the region'. They urged the Kenyan government to assume a more active role in helping to solve the problems of Somalia.

As the battles between the Ethiopian Garre and the Kenyan Ajuran intensified from early June 2000, those between the Ethiopian Boran and Gabra Miigo on the one hand and the Garre on the other intensified as well. On 26 June 2000, a meeting was held by Ethiopian government officials in Moyale. Representatives of the Gabra, Boran and Garre narrated their understandings of the causes of the conflict.

One Gabra elder accused the Garre of opportunistic handling of property issues:

> The Garre follow no custom. Be it the Islamic custom, the one of the Borana, or the one of the Government, they do not follow any of them. If we settle together in our land, they say that the land belongs to God. If our people settle among them, they say it is ours, of the Garre.
>
> They refused pasture and water to us, they eat alone. Recently they have taken a hundred *birr* from us if we water our animals within their boundaries. In the town, we can no longer sleep. They do not walk ahead of us, they do not follow us. Their way of life is that of the Somali, disorderly.

One Boran elder even accused the Garre of intentionally creating a situation in which they may benefit from refugee status: 'in the life of the Garre there is something of the way of refugees. They bring wars to this land and later become refugees, they seek livelihood in the hands of the Government.' Faced with the allegations made against the Garre community, a Garre elder responded: 'What we are accused of, is all lies. If we spoil anything which belongs to the Gabra and the

---

13. The RRA operates in the Bay and Bakool regions of southern Somalia between the Rivers Juba and Shabelle.

Plate 11.3 Melbana well, Ethiopia, 1986: horses come to drink from a trough that has been filled for the cattle. As former mounts of cavalry and messengers among the Boran, horses still enjoy privileges (photo: G. Schlee)

Boran, they themselves can ask us about it without taking recourse to the government. But we want to stay in peace with all of them.'

It is obvious that the conflicting parties were competing for the sympathy of the government and its support. This round seems to have been won by the Gabra and Boran. In response to all, a senior government official concluded that the problem was one of leadership. If there are OLF activities among the Boran, the government summons their leader, the *Abba Gada*, and holds him responsible. Also the Gabra have a *Talia* (title of an office-holder), Hassan Qalla, who commented:

But with you, the Garre, you have no leadership. You have no head. Every individual is his own leader. The young do not obey the elders. You have no defined culture which binds you together or binds you either to Boran customs of this land or to the government rule of law. We have repeatedly advised you to desist from warlike habits. It is now up to the government to take action against those found breaching peace. This country is not like some other countries where one causes chaos and gets away with it, the way some of you did in Somalia. You should live in peace and never deprive

---

14. The OPDO is the Oromo People's Democratic Organization, aligned to the EPRDF government.

others of their rights.

The above meeting did not stop the war. Following bitter clashes between the Ethiopian Boran/Gabra and Garre in the last weeks, where many people were killed and livestock looted, the President of the Oromia State, the OPDO[14] Chairman and the President of the Somali Region, who was at the same time Chairman of the oppositional Somali League, held a meeting in Moyale on 6 July 2000.[15]

The officials from Addis Ababa, after listening to the discussions, then concluded the meeting with a rather general appeal to all to maintain peace. They appear to have been briefed that the matter was all about an OLF invasion, backed by Kenyan Boran and Ajuran, and were surprised to find a multilayered conflict between the various communities within Ethiopia as well. They invited the elders to further meetings to solve these problems.

On 10 August 2000, it was reported by the Kenya Broadcasting Corporation that sixty-five Garre had been killed by Boran two days before in Ethiopia, at Arero, between Negelle and Yavello. Boran elders and Ethiopian government officers gave the number of dead as forty-eight, and that of the wounded as thirty-one. Not all of the victims were Garre. The area was one in which informal gold mining took place. Here Garre had killed four Guji Oromo days before. A combined Boran/Guji force then informed the Gabra, Boran, Burji and Guji to vacate the area. The Garre were not informed, and everyone who had disregarded the warning then came under fire.

The Ethiopian authorities started to consider a referendum as a means to stop the Boran/Garre conflict. The idea was that people in the south-eastern fringe of the Oromia state, between Moyale and Negelle, should vote on whether they wanted to this area to continue to belong to Oromia or to be transferred to the Somali Region. The Boran, however, found this unacceptable: they wanted the war to end first, so that everyone could return to their prior residence. Many Boran had been expelled from areas by the Garre; a referendum now would only legalize the results of the recent clashes. They also wanted their *Abba Gada* to be involved in the process. At the moment this was not possible because he was occupied with rituals that led to the handing over of his office to a new holder in January or February 2001. In August 2000, he was in the Arero region performing the *oda* ceremony, which involves sacrificial slaughters and prayers for the *hayyus* and other officials of the age-set that has newly moved into the *gada* grade. His own successor had been chosen and was moving around with him. Such ceremonies would later be held in different places all over Boranland. He had a heavy escort of Ethiopian government forces, a measure of the general level of insecurity.

Conclusion

The changing alliances beween pastoralist groups of northern Kenya cannot be explained only in terms of pastoralist conflicts, pastoralist systems, pastoral resources like pasture and water, and the like. While factors belonging to the sphere of the pastoral economy and ecology do have an effect on how alliances are shaped, factors

15. Interview by Abdullahi Shongolo with Ibrahim Nurow.

from outside also play a role. These include interests of urban elites as well as ethnic politics and administrative policies of the states involved, such as the delineation of district or regional boundaries.

Also the perceptions that state representatives have about what is going on at the 'tribal' level, including erroneous perceptions, have an impact on the course of the conflicts. These perceptions tend to be influenced by national and international security considerations. Warring states seek alliances with local or regional forces against each other, and even use third states as a battleground, exacerbating local situations. The result is a vicious circle of increasing conflict.

Religious identification is an important factor in mobilization but it runs into paradoxes. Rallying in the name of Islam makes little sense if the perceived enemies are Muslims as well.

Territorial claims, not just about pastures but also about political and administrative positions in a given territory, are put forward in ethnic terms. This is not new, but the undisguised way in which this is done may be new. In earlier periods, in my experience throughout the 1970s and 1980s, at least in Kenya but very probably also in Ethiopia, the ethnic dimension of politics was only apparent to insiders. It was talked about like a secret, although it may have been an open secret. More recently, ethnic politics have dropped all pretences to be anything else but ethnic politics.

Chapter 12

Roads to Nowhere: Nomadic Understandings of Space and Ethnicity

John C. Wood

Introduction

Is there a relationship between people's understandings of land and their sense of themselves as a society? Do nomads, who use places and spaces differently from settled populations, understand the dynamics, the limits, the boundaries of their social groups differently? And, if so, do these differences shape nomadic responses to ethnic and international others? This chapter asks these questions as it tries to make sense of the response – or lack of response – by Gabra nomads of northern Kenya to multinational petroleum exploration in Gabra grazing and watering areas. Contrary to my expectations, the Gabra did not see oil explorers as invaders or intruders. They did not always like what the explorers did, but it was their behaviour, not their presence, that annoyed. The chapter proposes that nomadic understandings of land – and ethnicity – emphasize nodes and trajectories rather than areas and boundaries, and that this difference helps us understand not only Gabra indifference to oil companies but also their relative equanimity to outsiders and change. It is not their borders that need protecting, but their centres.

New Roads in Gabra

When I began working with nomadic, camel-herding Gabra of East Africa, I often saw and sometimes used the straight dirt roads to nowhere that criss-crossed the rough desert landscape. People told me stories about white men and bulldozers that ploughed the roads in the late 1980s. They did not dwell on those years. And, while I shook my head at what I took to be a great insult to the land and to Gabra, I asked little about the roads or what they thought of them. My research focused on other matters.

Several years later, when I returned, I learned that the Gabra were suffering an alarming rise in cancer deaths. Cancer was not epidemic. Clinics here see hundreds, even thousands, of people annually, mostly for TB, malaria and stomach disorders. But doctors and nurses told me that the cancer rate at each of the clinics had leapt from no cases or one every several years to three to five cancer-related deaths a year, and involved types of cancer (throat, stomach and liver) that they had seldom, if ever, seen before. Gabra themselves did not know the disease as cancer but as *luqub lubu*, 'sickness of the throat', and they, who have good memories for such things, said it was *wan hereti* – 'something new'.

Gabra had multiple hypotheses about its cause: new veterinary medicines in the milk and meat they ate, drought-relief maize and other non-traditional foods, chewing tobacco, hot tea. But to my ears the most plausible was that the oil companies, which made the roads and drilled test wells through the fragile desert aquifers, did something to pollute groundwater. Worldlier Gabra – ones who had been to school and read the dated international magazines, such as *Newsweek* and *Time*, that were sometimes available – even wondered if oil companies, or others masquerading as such, had buried toxic wastes from elsewhere. Fuelling this speculation was the sudden death in one day in January 1998 of hundreds of goats and sheep that drank water from a recently reopened borehole at Kargi, an event that linked misfortune with water and reminded everyone that a 12,000-foot-deep test oil well had been dug near there in the 1980s.[1]

The next year I returned once more to find out what Gabra remembered of the petroleum company work, and to see what my untrained epidemiological eyes could find out about possible causes of cancer.[2] As an anthropologist, I wanted to know whether and how the oil exploration had penetrated their consciousness, not simply as a discrete memory, but as a transformative experience, as something that had reshaped their understandings of themselves and others. Gabra had witnessed first-hand the technological muscle of the global market economy – huge bulldozers, helicopters, drills, seismic trucks, cables, computers and satellite dishes. They had seen European and American engineers driving Land Cruisers everywhere at breakneck speeds and talking into satellite phones. They had seen tractor trailers as long as camel trains roll across the desert to deliver enough steel pipe for three test wells, each more than two miles deep. They had seen the region scored by roads in directions that seldom made practical sense to them. The roads were seismic cutlines. Huge trucks positioned on the lines shook the earth. Sensors recorded 'echoes' from the geological depths. Other trucks with computers gathered this information, processed it and made images that geologists could read and use to predict where there might be deep pockets of oil. Hundreds of Gabra men got jobs with these companies at various stages of the exploration, and the result of these jobs was income, which they invested either in animals or school fees for children or, in some cases, frittered away on luxuries and alcohol. Some animals bought with oil money were named after the companies – 'GSI' – or the pay – 'Noti'.

In my view, Gabra had been invaded by a colossal external power associated with Europeans (many of whom were actually Americans – the exploration was done for

---

1. It remains uncertain what killed the animals. GTZ, a German development organization, tested the water and found it highly alkaline. The well may have been polluted by run-off from the 'El Niño' rains the year before; the well had not been used since the rains.
2. I collected water and maize samples in May and June 2001 and brought these back to the United States for testing. So far nothing has been detected. The water was tested for arsenic, a growing groundwater carcinogen; the maize was given a preliminary test for aflatoxin. Tests are expensive and it helps to know what you're looking for, which at this point we do not.

Amoco, a multinational company) and down-country Kenyans. Land had been permanently altered. Lives of workers had taken sharp turns, not always for the better. I expected to find indelible traces of this 'invasion' scored into Gabra memories, just as I found seismic cutlines across the land. I thought oil exploration would be a watershed event in their history, that Gabra would feel violated and weakened by the foreign power over which they had no control. Wasn't oil exploration a new form of colonial occupation?

I was surprised to learn something quite different. Very few Gabra had felt invaded by the oilmen. Most said they were glad the explorers came. The ones who had got work said they would sign up again if oil companies returned. Many said they wished the oilmen had discovered oil; then, they thought, Gabra would be rich. Those who had not worked for the oil companies, and had little or no understanding of what they were up to – what oil even was – shrugged their shoulders with indifference. Sure, they remembered the equipment, they had watched the activity, but it was none of their business. The episode seemed not to have captured their imaginations. It hadn't piqued their interest the way a severe drought or livestock raid or a shipment of relief food did. Just about any Gabra you asked had stories about drought and famine. About oil exploration, I got little of the meaningful narrative I'd hoped to hear. Gabra, who name years after salient events, named no year after the oil companies. The more I asked them about the possible sources of cancer, the more I learned that the oil work was not a central hypothesis at all, but one of many, all of which had to do with recent events or new practices: the cancer was new, so its cause must be new as well, whatever it was.

What interests me is the disjunction between my sense of what Gabra ought to have thought about the oil exploration and what they seem to have thought. When I lived with Luhya farmers in western Kenya in the mid-1980s, I heard my hosts talk bitterly – or at least warily – about the British occupation. I heard similar sorts of things from Kikuyu in the central highlands. In contrast, Gabra elders who remembered the colonial era spoke of it as a relatively good time. Some even asked hopefully if I knew when the British might return, for they said when the British returned good times would return as well. Of course, these differences might be explained by their different experiences of colonialism: the British occupied land and displaced Africans in Kenya's central highlands and western plateau. In the 'northern frontier', the British simply sealed off the place and left people more or less where they were, and, while they set some irritating restrictions on what people could do and where they could move, Gabra said the British kept the peace and that made it a prosperous time for them.

I thought the oil exploration in the north was an intrusion much more analogous to the British occupation of the central and western highlands a century before. It threatened displacement. It changed the landscape, probably for ever. Why weren't Gabra more exercised about it?

There is much scholarly interest these days in modernity, globalization and changing ethnic identifications. We are keen to clarify what people do with 'outside' influence, whether in resistance or acceptance or both. We want to know how syncretism, hybridity or change works. But we might also wonder whether 'change', which seems a prime concern of current ethnography, is the right concept: at least with respect to nomads. Does Proteus change being Proteus when Proteus changes form? Yes, he changes, but not in the usual sense of the word: he remains Proteus not despite but because he changes; it is his nature to change. A nomad's change is a different sort of thing from, say, that of a settled farmer, who lives in the same house, in the same spot, and tills the same dirt, milks the same cow, year after year, and his son the same. If the nomadic self and social group are not singular and fixed but always, and by definition, complex sets of moving parts, then are we sure that 'change' is the appropriate term for the processes we call change? Perhaps it would be better to imagine that people are improvising with forms more or less available from the beginning, deploying some in one circumstance, saving others for another (see Galaty 1993: 177).

Perhaps different cultural traditions 'change' in different sorts of ways. Perhaps Gabra indifference to oil exploration has to do with their being nomads.

Whose Models?

These musings are aimed at the following point: change of identity might happen to different sorts of people in different ways, according to their more basic understandings of person, place and society, and we who study change ought to pay attention to such differences. Thus there are two questions. Do nomads, who use space differently from settled people, also think of space differently? And, if so, do nomads think about the dynamics, the limits, the boundaries of their social identities differently?

Western anthropologists, whose business is making sense of cultural differences, often use metaphors of space, presumably borrowed from our own culture, to make sense of other people's identities. We have all heard such language. We've probably all used it. People 'shift' identities. They 'locate' themselves in one ethnicity or another. They 'inhabit' identities like tents or houses or countries. Their identities have hard or soft 'edges'. They 'move', especially in these post-colonial times, from one 'location' to another, sometimes 'through' transitional 'spaces', often across 'boundaries'. Indeed, boundary metaphors, at least since Barth's *Ethnic Groups and Boundaries*, are especially pervasive. In that text, creating and maintaining boundaries are what ethnicity is all about. Barth's important point was that boundaries did not so much divide as relate different groups. 'In other words, ethnic distinctions do not depend on an absence of social interaction and acceptance, but are quite to the contrary often the very foundation on which embracing social systems are built' (Barth 1969: 10).

Whence came the space metaphor? Why 'boundaries'?

It is in fact difficult to talk about identity – any sort of identity: cultural, personal, class, ethnic – without making reference to space and harder still to talk about cultural change without referring to movements across space. In fact, it is well understood that spatial logics serve as a basis for understanding many different domains of human experience, from abstract algebra to social classes. The connection between space and identity is easily and intuitively grasped. Identity is always, or at least usually, formed in some sort of place, in space and time: it is next to impossible to have identity – that is, a sense of sameness or similarity with something else – without space. As persons, we come to our identity in space; identity changes in our movements through space. Rites of passage make use of spatial shifts to mark identity shifts (van Gennep 1960; Turner 1995). Identities accompany our movement across space, or space accompanies our roving identities. Whichever way the relationship flows, there seems to be a relationship.

What is less clear is whether people from different cultures move through the same or different spaces. Clearly different sorts of people inhabit different places. Their landscapes and the names of landmarks and suchlike are not the same (Casey 1996). What I wonder is whether different people think of and use the dimensions of their space in the same or different sorts of ways. There is strong evidence that they do at least speak about space differently (Levinson 1998). Is everybody's sense of ordinary space equally Euclidian? Does every society's idea of space involve (or emphasize) area? Do all spaces have edges, borders, perimeters, limits, lengths, widths, measure? Are all of these characteristics of space salient in the same ways? Are other aspects of space emphasized?[3] And, if so, to invoke a time-honoured question in cultural anthropology, do we err in using our spatial metaphors to talk about their identities if their logics of space are different from our own?

Ethnic Boundaries

Even when Western scholars critique the metaphor of ethnic 'boundaries', our intent has usually been to clarify and refine, to push the idea in new directions, to encompass new aspects of what ethnicity might involve (a play for tactical advantage, a socio-aesthetic practice, an aspect of imaginative consciousness). But no one that I am aware of wonders about the implications of spatial metaphors in the first place, or considers the possibility that just as ethnicity differs from society to society, indeed, within the same society, so might people's basic understandings of space, and with it the basic structure of their identities.

Cohen (1994), in a collection on ethnicity subtitled 'Beyond "Ethnic Groups and Boundaries"', wisely wonders whether anthropology has thought enough about what ethnic boundaries mean. He thinks not, and he suggests that boundaries are often convenient fictions, which work better for observer than observed. '[A]nthropologists', he writes, 'have been largely content to assume the existence and integrity of collective boundaries ... Rather than questioning their existence, or questioning the extent to which they might reasonably be generalised (whose boundaries are they?), they have been concerned almost exclusively with the ways in which boundaries are marked' (Cohen 1994: 64). Despite his critique, Cohen doesn't reject 'boundary' as a metaphor but proposes instead to retune its meaning. He wants us to think about how individuals come to know and experience their boundaries, rather than seeing boundaries always as merely strategic resources. Too much

---

3. I wonder in fact whether space is like two other culturally variable concepts: gender and race. All three have been assumed to be fixed physical realities, rather than constructions understood and elaborated by different people differently.

emphasis, he writes, has been put on the 'lines which mark the extent of contiguous societies' and not enough on lines between persons across or even within groups, or for that matter on personal boundaries crossed during the life course (Cohen 1994: 65). Cohen calls our attention to subtle transformations within the person, within the community, between communities. Yet, even as Cohen would redefine 'boundary', he does not propose to abandon the concept. In the end, he asserts, 'boundary-crossing stimulates the individual's self-reflexivity' (Cohen 1994: 69). Boundaries, he says, are 'zones of reflection: on who one is; on who others are' (Cohen 1994: 74).

Galaty, in another rich and complex discussion of ethnicity, nevertheless uses the boundary term in ways that get in the way of the sort of processual analysis he is otherwise developing. 'Boundaries', he writes, 'can be collapsed as well as constructed, can experience decay or be maintained. This dynamic aspect of identity, witnessed by residential shifts across boundaries and longer-term shifts of boundaries and marked by multilingualism and code shifting, has led some scholars to suggest that ethnicity in pastoral East Africa ... tends to be "fluid" or "mutable"' (Galaty 1993: 176–77). Galaty argues that 'ethnic "shifters"', which people can pick up and lay down as more or less familiar tropes or signs, enable all of this boundary crossing.

Again I wonder whether 'boundary' is the right metaphor. Galaty writes: 'Their projects surely include straddling a boundary' (1993: 177). But in what sense? How does 'boundary' help us understand what is going on when people maintain relationships with different sorts of people? The language seems to presume a sort of transgression. The 'problem' of all this 'boundary shifting' is a problem, in part, because it seems 'shifty'. If, on the other hand, we assume that people are involved in complex relational networks, our interest shifts from their 'shiftiness' to their enterprise, from crossing to relating.

I commend the spirit of Cohen and Galaty's efforts to move to more experiential and nuanced understandings of ethnic identities. But I fear that preserving the concept of 'boundaries', without more reflection on its implications – on how it is bound up in cultural relativities – leads us back to the very assumptions about identity that Cohen and Galaty (and Barth before them) are trying to avoid. They each begin their essays with the observation that the impetus behind the use of the terms 'ethnicity' and 'boundaries' was an interest in cultural processes over static structures. Odd then that the 'boundary' metaphor has stuck in everyone's imagination. For, however mutable, a boundary is basically a line, a static structure, a thing to be respected, crossed or straddled.

Loose Identities

Nowhere in my work with Gabra nomads has the difference between us been more apparent than in our different sensibilities about space and identity. Their sense of space, compared with my own, seemed decidedly non-Euclidian. Near (*dio*) was often very far (*fago*), or so it seemed to me, and far was sometimes just nearby. Likewise, inside (*olla*) – that is, a socially and morally central location – was in some sense outside (*ala*) relative to some other centre, and in that sense outside was also inside. Space was far more a social, moral and cultural construction than I had ever thought, despite my having read Evans-Pritchard's (1940) celebrated discussion of just this very thing. Gabra spatial understandings emerge from their nomadism. Nomads move, shifting from one location to another, never in ordinary circumstances to the same place twice. Each time they pitch their tents anew, they recreate the old place (Deleuze and Guattari 1987). They are always moving and in some sense staying put. Their movements constantly domesticate wild space and return domestic space to wilderness. What seemed a contradiction to me was in fact nearer their everyday experience.

Gabra had similarly complicated ideas about their own and their neighbours' identities. Indeed, the fact of changeable ethnicity among East African pastoralists has been widely noted and discussed (see *inter alia* Sobania 1979, 1980, 1988b; Schlee 1994a [1989]; Spear and Waller 1993). It is also well known that colonial governments in Africa fixed identities that had been at least flexible, and may never have existed. It is thus curious that scholars have continued to study 'ethnic' communities as if they were the appropriate units of analysis. So strong are some Gabra ties outside Gabra and so weak are some ties among Gabra that I have often wondered whether studying them as a 'single' ethnic community makes any sense; yet have gone ahead and done so.

Perhaps, as Aguilar (1999) has said of the Waso Boran, anthropologists have been too quick to play the 'ethnicity' card, to define what on the ground, in daily life, is diaphanous and gauzy. In actual relations, people do business with each other, compete, share, love, revile, see through duplicities, accomplish deceptions, put up with or expel in ways that are far too complex to reduce to a label. Just as the 'boundary' metaphor stumbles when it tries to describe what happens, say, between Gabra and their nomadic (or settled) others, perhaps 'ethnicity' also falls on mushy ground. Perhaps ethnicity is something that happens only at certain times, at certain levels: say, in relations with the state. 'It is very difficult', Aguilar writes, 'to assume that ethnicity is created in a daily level of social relations' (1999: 150). People work hard on their identities. It is not always clear that they think much about ethnicities.

Changing Identities

A Gabra friend, Tumal Orto, told me how his father, back in the 1930s, met an Indian trader who changed his life. 'My father', Tumal said, 'was a pastoralist man, and on top of that he had a knowledge of business.' Orto, Tumal's father, had had to come to his knowledge of business. This is the story. In those days, Orto kept his goats and sheep in the Waso Nyiro area near Garba Tula east of Isiolo, in the southern part of what the British then regarded as the Kenya colony's 'northern frontier'. The region is a flat expanse, shimmering with heat, covered with thorn bush, sand and rocks, broken occasionally by isolated volcanic outcrops in the shape of pyramids. Most of it is too dry to farm but it provides good pasture for animals.

Nur Mohammed, an Indian trader, was working for a businessman in Meru, to the south, but Nur wanted to strike out on his own. Knowing there was an untapped market among pastoralists beyond Isiolo, he loaded a string of donkeys with tobacco, sugar, sweets, tea, cloth, blankets, pots and cooking utensils, and clanked off into the desert. 'There was a wild rhino,' Tumal said, recalling a story he heard countless times as a child:

There used to be a lot of rhino in that area. The rhino jumped over him and knocked him down. Just while he was driving the donkeys. People heard and raised an alarm. My father was among those who came and they carried Nur to my father's camp. My father slaughtered a sheep, made a soup for him. Nur stayed in my father's camp for some weeks while he recovered. It was during this time that they became friends.

The two men, from different worlds, talked of herding and business and the future and, during those weeks, developed an enduring liking and trust for each other – to this day their sons maintain friendly business ties. The relationship was greased by Nur's already knowing the Boran language, the mother tongue of Gabra and the lingua franca of Kenya's north.

After his recovery, Nur began to market Orto's sheep and goats in Meru. They both made money. Nur turned his profits into a shop in Isiolo, where he apparently prospered selling war rations to the British. Orto bought more goats and sheep. Eventually Orto moved his family north to the slopes of Marsabit, about halfway between Isiolo and Moyale on Kenya's Ethiopian border. Orto was now a businessman. But he had not abandoned the desert, given up caring for animals or quitted his place of birth. He had changed – and this is the important thing, a central feature of nomadic culture – by staying the same. Alteration was an essential part of who he was. It is what Gabra men do: go out, leave home, venture away from the centre, make friendships, have exotic experiences, and then return transformed by that knowledge (Wood 1999: 113–14). To be a nomad is to move, and to move is to change, so for a nomad, as for Proteus, to change is, in some sense, to stay the same.

'My father was using new techniques,' Tumal said. 'He doesn't take chai without sugar now, because he has learned. He cannot take tea without sugar. He's always having cash. He puts on good clothes. He buys, sometimes coming to town with a camel, foodstuffs, rice, a bag of rice, a bag of sugar, some tobacco, for home use and for sale.' Saying 'techniques', Tumal emphasized that his father had picked up new skills, like learning to tie a new knot or hum a strange tune. But it isn't clear that anything at the level of the identity had changed. Orto, businessman and ally of Indian merchant Nur Mohammed, remained a Gabra, an Odola, a desert herder.

Tumal told the story about his father by way of explaining his own life. Tumal is a shopkeeper, not unlike Nur, but in Marsabit town. Like his father, he is rich in goats and sheep and camels. Though members of his extended family remain nomadic pastoralists, who shift with their animals and keep their herds for subsistence and symbolic value, Tumal and his brothers now live mostly sedentary lives and raise livestock almost exclusively for sale (though they continue to take a pastoralist's pride in their ownership of livestock). They are businessmen, turned that way by their father. Tumal is also a Gabra, a member of a camel-herding community with historical ties to Boran, who are cattle-keepers in the surrounding highlands. But Tumal's father, Orto, was born a Rendille, a Cushitic-speaking community of camel herders south of Gabra. Orto became a Gabra as an adult. Depending on whom Tumal was talking to, he has identified himself as Rendille, Boran, Gabra, Odola (a clan shared by both Rendille and Gabra), Kenyan and businessman.

Tumal's story has certain particularities, as does anyone's story, but talk to anybody in the north of Kenya or the south of Ethiopia about the past and you quickly get the sense that identity-wise things have been pretty mixed up for a long time – at least mixed up by our sedentary standards (for the long view, see Schlee 1994a [1989]). When I began reading about East African pastoralists generally and studying Gabra in particular, I puzzled over the meaning of ethnic identity in the region. I wondered about people's capacity to change identity, to reach out and take up new ways of life, to embrace others, to make all sorts of new alliances and associations, even with enemies, and yet – and this is what piqued my sedentary sensibilities – sense no contradiction. Asked, they shrugged their shoulders and explained these shifts and ties as they would explain marriages between families. The sense that things were mixed up and contradictory was clearly my own.

What interests me is the lack of boundaries, or at least the lack of emphasis on boundaries. Nur needed help; Orto took him in. A lasting friendship – an alliance, if you will – grew between them. Orto easily slipped into a new way of managing goats and sheep – managing them for sale to an external market. He changed his economic niche, without also changing his ethnicity, for he and his family would have thought it strange to say of him that he was no longer Gabra because he was selling goats and sheep rather than raising them for subsistence and social capital – nowadays, most Gabra sell goats and sheep for money. On the other hand, Orto had changed his ethnic identity, from Rendille to Gabra, without changing his lifestyle or economic speciality: he remained a nomad who lived in a matted tent and followed camels, goats and sheep through the scrubland. His material life, what he did in his days and nights, stayed much the same as before. His field of friends expanded to include northern Gabra pastoralists, augmenting his birth ties to southern Rendille. He remained Odola, a member of a clan shared by both ethnic groups. The idea of crossing a boundary simply does not help us to understand Orto's life. What was crossed when he became a businessman? Nothing. Social and economic worlds expanded. On the other hand, when he changed from being Rendille to Gabra, his life, his daily work, his material surroundings, did not really change; they stayed to all intents and purposes the same. So, where he changed his life, he did not think of himself as changing his ethnicity. And, where he changed his ethnicity, he did not think of himself as changing his life.

Source Domains

Certainly Westerners are used to using metaphors of space to describe the people we study. We draw on ideas about land to speak of 'ethnic groups and boundaries'. We think of Western values intruding on indigenous people. We write of Boranland or Gujiland, even if the people we describe don't think of land that way, or at least not with the same connotations. We talk about groups migrating from one identification to another, of one ethnic group joining another, of cultural movements and encounters – all of which we envision in spatial terms. The fact that we use metaphors to make sense of complex ideas is basic to human cognition: we draw on familiar, concrete experience to make sense of unfamiliar, abstract conceptions (see Lakoff and Johnson 1980; Johnson 1987). This is consistent with Fabian's related perception that anthropological knowledge is ultimately derived from shared experience with informants (Fabian 1999). The problem with cross-cultural metaphorical thinking is that we risk mixing our metaphors: seeing 'the other' merely as a projection of ourselves. What if our spatial metaphors are inappropriate? What if the people we describe do not understand themselves to be crossing boundaries, do not think of others as intruders or of alliances as migrations or joinings or movements or encounters, but use other metaphors? Perhaps their understandings are not predicated on space, or their understandings of space are different from our own.

A farmer's wealth is in land; land is what captures the farmer's imagination, fills the farmer's dreams, informs the farmer's thinking about other things. A pastoralist's wealth, at least a Gabra's wealth, is in animals – and it is more likely herd dynamics rather than space per se that captures the herder's interest. Except in the loosest way, Gabra do not talk about camels (or themselves) being tied to particular land. Gabraland, *lafti Gabraa*, is where Gabra are (see Schlee 1990a: 24; and below). They don't talk about places themselves as *lafti Gabraa*. When, for example, urban Gabra speak wistfully of Gabraland they are referring to themselves as estranged not from land but from people, their community.

Gabra Spaces

Elsewhere I have written about the symbolic inversion of Gabra men becoming women, and linked this inversion, in part, to Gabra use of space (Wood 1999: 185–89). In a word, young Gabra men leave the social centres, the main camps, for satellite camps in the dangerous distance to tend their family's animals. Later, as mature men, they marry and begin to split their interests between the pastures and camps. Finally, as old men, they return full-time to the main camps to take charge of life-sustaining rituals, blessing, and prayers. It is in this last phase that they are women.

Just as men who are women are not completely women, having been men and continuing to be men in many respects, the spatial opposition between inside and outside, camps and pastures, is not simple either. The Gabra distinction between inside and outside is at least partly moral. Thus, to speak of inside is to refer to the society, the community and its core values, while to speak of outside is to think of activity beyond the reach of strong social order. Yet outside is also the location, the home, of the herds of camels, goats and sheep, which are important not only economically but also morally. Indeed, as I mentioned above, outside is also inside, and inside is also in some sense outside (Kassam and Gemetchu Megerssa 1994; Wood 1999). These polar opposites are caught up in each other, are not really opposed at all, and, sharing aspects of their opposite, capture the spatial ambiguities Gabra experience as nomadic pastoralists.

Spatial ambiguity is also reflected in ritual performances, which normally occur at main camps. The most important rites of passage, however, the *jila galaani*, which mark the transition of Gabra into new grades of responsibility, take place only after a lengthy pilgrimage to the farthest outside, an outside in the dangerous distance, but also a point of mythic Gabra origin, an ur-home, an inside of insides. On the way to these rites, Gabra and all of their animals pass through a series of thresholds, or gateways, placed on the bushy plains between extinct volcanic mountains. These doors are composed of cut branches of trees as well as representative individuals holding long open bolts of cloth, creating an aisle of sorts, a ritual gauntlet. Upon doing this, they 'enter' what van Gennep and Turner have called 'liminality'. It is during the following weeks that the key rites of passage occur. But, though they pass through a doorway, they are not so much moving from one space to another as from one time to another. Indeed, participants must leave impure items – such as goods bought in towns in the market economy – behind. They may not pass through the gates with these goods. But, once on the other side, the owners may go around the doorway, pick up the profane objects and continue on with the rest to the ritual grounds (G. Schlee, personal communication). The one side is the same as the other – there are no sides, there is no boundary. There is just a doorway, a passage through not space so much as time or status.

Rendille Spaces

Rendille, who are neighbours to the south of Gabra and share similar material and ritual culture, also seem to share similar ideas about space. Schlee, in a paper about pastoral participation in national markets, indicates that Rendille express 'a rather strange lack of two-dimensional representations, or surface areas, especially of bounded surface areas', and 'often seem to lack the possessive feelings about "land"'(1990a: 7; see also Schlee 1998a). To illustrate, Schlee (1990a: 23–24) recounts an interview with a Rendille elder named Barowa, who expressed views about space that I immediately recognized from my own interviews with Gabra elders.

'When you say "that is Rendilleland," "that is the earth of Rendille," where is that?' Schlee asked the elder. Barowa reacted with an enumeration of localities where Rendille can be found now, not with a description of the course of boundaries. Some of those places were in areas typically regarded as 'belonging' to other ethnic groups, but into which Rendille had taken their livestock. Barowa said: 'The only boundary [mpaka – a Swahili word that has been introduced to the area by the colonial administration] there is, is the one of fear,' i.e. the Rendille can take their animals as far as they dare (Schlee 1990a: 23–34).

Schlee concludes his account of the interview with the observation that 'Rendilleland' is 'wherever this settlement cluster happened to be'. He goes on to make the important distinction between two-dimensional spaces, as farmers maintain with their cultivated fields, and zero- and one-dimensional spatialities – points and trajectories – as nomads emphasize – wells and holy grounds and encampments being points of significance, and paths between pastures and wells, camps and holy grounds being trajectories of significance. He observes that, much like Gabra, Rendille nomads have a strange lack of boundaries, perimeters, lines between spaces.

Roads to Nowhere

Nomads use and talk about space differently from most of us who are settled. And, in so far as their identities are caught up in wider contexts, including their place and its logic of space, they also think about themselves and their relationship to others differently. In May 2000, I travelled from Torbi to Bubissa and back to Marsabit, thence to Maikona and along the Chalbi rim to North Horr, and up to Balessa and Dukana, all the while talking to Gabra about the current drought and cancer. The next year I travelled from Marsabit to Maikona, Kalacha, North Horr and up to Balessa, then back to Maikona, via Kalacha, and across the desert to Kargi. This time I spoke with clinic workers about cancer and with elders about the oil exploration. During the trip I alternated time between nomadic and settled communities. I was interested in Gabra memories of what the oil explorers had done. I found myself hearing much the same thing from place to place.

Most people I spoke to either shrugged their shoulders in indifference – the oil companies came and went a bit like the weather, as something over which they had no control – or they said the oil exploration might have been a good thing for Gabra. If Gabra were nonplussed by the presence of the oil companies, they were certainly aware of their presence, attentive to their habits and activities. They were curious, but in the way they would be of a strange animal that wandered past where they were living. It is hard to indicate this distinction, this mix of interest and indifference, this untroubled awareness, except with quotations.

Elema Konchora at Balessa remembered the day he came upon the oil people working at Kilkile, where they dug a test well. He and his companions were taking care of goats and sheep.

'What's happening?' they asked themselves. 'What are these people doing? Why are they doing it here? Who are they?'

I asked Elema what he thought of their activities. At this point I was still interested in uncovering Gabra resentment or resistance. I wanted very much to find out that Gabra had felt exploited, violated by the big oil companies and the 'modern' world. But that was not what I heard.

'They had their own business,' Elema said, 'and we had our own business, and we had nothing to do with each other.'

What was their business? Elema described accurately and at length the activities of the oil companies, the building of the seismic roads, the laying down of cable, the vibrating trucks. He spoke of the oil drilling equipment. But was he not angry that they had dug up the land?

'No,' he said. 'People gave little thought to the GSI people.[4] We did not concentrate on those people.'

Disbelieving this, I pointed out that, had the strangers been Dassanetch or Rendille, they would have paid attention, they would have wondered what they were doing and they might have wanted them to leave.

'True,' he said. 'But these people [the oil companies], we did not regard them as enemies.'

OK, I said. But, if the strangers had been Gabra, people surely would have talked about them, wondered about their business. They would have wanted their news.

---

4. GSI managed the seismic research. Some Gabra used it as a gloss for all the different companies, including AMOCO, which was in charge. Some also spoke of 'Geosource', another contractor, and of 'Kenya Oil', which was supposed to be partnered with AMOCO.

'They [the oilmen] were in between us,' he said. 'Not enemies. Not part of Gabra. They were somewhere in between [*gidu kes*].'

In fact, many Gabra I spoke with seemed troubled, not by the presence of the oil companies in the region, but by their aloofness, their indifference, their social distance from Gabra. 'They were quite isolated, and one wonders why,' said Adele Tura, a Gabra and corporate executive in Nairobi, who remembered the oil exploration from his school days.

The oil companies had strict rules of separation between their worksites and Gabra communities – presumably to protect people from getting hurt, or at least keep them out of the way, but in any case an antisocial separation that Gabra did not understand. There was also separation between types of workers at the work camps. All of the camps kept 'experts', who were mostly white, separate from 'nationals', who were black. Even Kenyan geologists and engineers who were 'experts' were reportedly segregated from European and American specialists. Drivers would not give lifts to Gabra on foot nor stop to offer them water or even to pass the time. These separations were noted by Gabra. 'Why weren't the oil people friendlier?' they wondered.

They were bothered by other things the oil companies did. The oil people moved their camps on inauspicious days of the week, and Gabra who worked for them had to shift with them on those days. The explorers refused to use their oil-drilling machinary to dig wells for water, which would have been useful to pastoralists. When they left, they took everything with them, or buried it, but did not give things to people, even their garbage, some of which would have been useful. It was an affront, what they did and did not do. But it was not an affront their being there.

Isaako Godana at El Gade observed that people at first feared the oil explorers and then thought they might be a good thing:

> There were many people, many vehicles, and they made many roads. They made the roads to find the point where there was oil, to find out where to drill. They never consulted the community. They did not sit with elders and give them the nature of their work. When they first appeared we were fearing. What is happening? When we heard that they were looking for oil we thought that this might be something that would benefit the community.

Gabra have beliefs about life-giving or life-sustaining aspects of soil – *koshe* – and I wondered whether they thought all the digging in the soil would disturb the *koshe*. None I spoke to did. The nearest I heard to a complaint about the oil exploration's effects on land was from several elders who worried about the seismic cutlines, which served as roads, albeit not always running in the directions one would like. They feared the roads would give outsiders – enemies as well as police and army officials – easier access to them. Gabra dwell in an area strewn with lava boulders. The rubble covers the landscape and makes vehicular travel, even with supposedly all-terrain vehicles, nearly impossible. The only way to travel over much of this terrain is on foot. Gabra are used to it. They are experts at walking long distances over the stones. Most down-country Kenyans and even some of Gabra's pastoral neighbours are unwilling or unable to walk for long-distances over such rugged land. Gabra felt safe on the stones; the roads threatened their sense of safety. But even elders who said they had first disliked the roads grew to like them. Gabra have taken to following the cutlines, even when they take them slightly out of their way, because travel on them is so much easier. Moreover, it is possible to get vehicles closer to some areas than ever before; on several occasions sick people have been evacuated by car where before they surely would have died. 'In fact,' said Robale Guyo, the Gabra chief of Forole, 'people are happy because they are getting so many roads, those cutlines that their camels are using.'

Besides the roads, were there other benefits from the oil companies? 'The outcome was very little,' said one man at El Gade. 'There was the well with water. Even that one it was not the petrol company that dug it. They [the oil company] just told where there was water. There was the benefit of employment. Many people worked for them. Except for those things, there was no benefit.'

Of course, a few Gabra did see the exploration as an encroachment, an invasion. These were men who had been to university and saw the petroleum companies in league with a hostile and indifferent government. They were not worried that the oil people had trodden on Gabra land; they were bothered that they would exploit Gabra powerlessness. 'Our people', said one young man at Marsabit, 'see the government as omnipotent, like a god or something. They have no spirit of resistance. They just lie down in the face of oppression ... This is all part of the same thing. We are ruled by this government from down Kenya. And they [the government officials] don't recognize us.'

Kenya's then President Moi flew to Kargi in 1987 after a test well there produced promising signs of oil and gave a speech about the great potential of oil exploration. The whole enterprise was thereafter linked with government. Gabra I spoke to felt they had no say in the matter one way or the other. One shopkeeper at Marsabit said, 'Our people do not have all that much capacity to refuse. It is from the top, you see.'

By far the most common response was no response at all. 'I've not worked for them,' a man at Balessa said. 'I've not understood their language. They don't understand mine. How do I know what they were doing?' He did not regard any of what the oil companies did as worth discussing.

Towards the end of our Chalbi tour in search of memories about the oil exploration, my Gabra companions and I heard news on the radio that Kenya had signed a deal with a consortium of Russian oil interests to do more oil exploration in northern Kenya. The radio reports did not indicate where they would be working. Newspaper reports I read later in Nairobi suggested the work would be east of Gabra areas. But as we listened to the radio I was curious to hear what my friends thought of the news: 'We don't know what they [the Russians] will bring,' said Mamo Guyo, who had worked with GSI as a driver, 'but the first priority is jobs.'

Conclusion: Senses of Space and Identity

Metaphors of 'territory,' 'area', and 'boundary', drawn as they are from sedentary experience, do not adequately capture Gabra senses of space or identity. Land is not something people here have owned. If others' activities are not in conflict with Gabra activities, others may move across and dwell on land occupied by Gabra.[5] Gabra do not identify with land and region in the ways sedentary Westerners do. Gabra do not have territory with boundaries that define where they may go or, by extension, who they are.

On the other hand, if Gabra were open, or at least indifferent, to oil exploration, they are not always or absolutely indifferent to others. It is not as if they make no distinctions between themselves and others, or that they do not seek to protect their Gabraness. Nor is it that they have an undifferentiated, unvariegated sense of land and space. Places matter. Earlier I mentioned the *jila galaani* pilgrimage – a ritual return to a mythic place of origin – an ur-centre – in this case, a mountain. The pilgrimage, held roughly every fourteen or twenty-one years, involves a series of rites that organize the passage of men from one status to another. The pilgrimage, like other rites such as marriage and *sorio* sacrifices, is also defining: Gabra see its performance as shoring up Gabra against various processes of decay and dissolution.[6] Gabra are quite particular about who may participate. At some rituals, persons who are not Gabra, who are not Gabra in good standing or who are not present with all relevant kin may not participate in or even witness the events. The spirit of these rites seems to be to instantiate, and thereby define, an ideal type of Gabra.

The *jila* pilgrimage ends at a sacred mountain, on the top of which, for the Galbo phratry, dwells a fantastic dragon-like monster, part snake, part lion. Only a small group of clan representatives climbs to the top, and, in the end, only one ever sees the creature. He feeds it fat from a sacrificial animal, and then returns to the plains below to report on its well-being. For the Galbo, the mountain is Forole, on the Kenya-Ethiopia border, and it is to this mountain that some trace their clan origin: the mountain was formed in mythic time out of the mist that protected one of a pair of brothers, presumably of divine origin, the other of whom was captured and grew up to found a Gabra-Galbo lineage. Schlee (1990b) has discussed the importance of the mountain as a sacred and ecological reserve: nothing on it may be removed, broken or sullied.

If Gabra do not have territories in the sense that most of us sedentary people do, they do have critical centres, such as Forole and, to a greater or lesser extent, wells, springs, other sacred mountains and shifting nomadic encampments. These centres are defining: they locate, orient and identify Gabra in their nomadic comings and goings. They do not have boundaries in the conventional sense of that word, and the use of that term mystifies the real significance in nomadic life of centres and pathways to and from centres, of nodes and trajectories.

---

5. Of course, individuals and institutions more powerful than Gabra, such as those with the support of the state, are able to move across and dwell in these areas, regardless of what Gabra think, which fact adds another wrinkle largely unexplored in this chapter. It could be that spatial logics are also drawn in part from people's relative capacity to resist intrusion. Even so, this confirms my thesis that distinct, relative experiences shape local sensibilities about space and identity.
6. Pilgrimages are organized within the five Gabra phratries, or mega-clans, and are specific to each. Their performance, however, is of great interest to Gabra generally, and even those who do not go draw comfort and satisfaction from their occurrence.

Gabra are open to others, even outsiders such as the oil companies, but not absolutely so: they are open to others who accept – or at least do not contradict – Gabra beliefs and practices. Ask just about any Gabra man his lineage and you'll hear names – fathers, grandfathers, great-grandfathers, and so on, back ten or even twenty generations. Often these names belong to other ethnic groups: Samburu, Rendille or Somali usually, but sometimes others. Such acceptance of ethnic difference is masked or confused if we describe Gabra relationships with others as occurring across boundaries. Indeed, this discussion indicates that our usual models of identity do not apply everywhere the same. Perhaps we should borrow metaphors from nomads themselves. If ethnic 'boundaries' are not appropriate because under-cognized, then 'gateways' might do. A doorway, more than a boundary, stresses the possible – though not absolute – openness between communities, and carries with it the sense of agency involved in opening and closing the gate. And it is a spatial metaphor nomads such as Gabra recognize and use themselves.[7]

Concepts of space are not everywhere the same, or at least have aspects that are not emphasized everywhere the same way. Judging by Gabra materials, there also seem to be rich and as yet unexplored connections between space and identity, such that people with different understandings of space may also have different understandings of identity. As anthropologists, we know better than to assume that our models of space and identity apply cross-culturally. Concepts of 'space' and 'identity', like 'race', 'sex', 'gender', and 'kinship' – once thought to persist more or less evenly and unproblematically across cultures – do seem in fact to be culture-bound. If we are to understand the changing relations between ethnic groups and changing ethnic identifications, we had better grasp not only the varying details but also their underlying structures. Sedentary models, at least when they are applied to nomads, lead us nowhere.

---

7. In fact, Gabra refer to different Gabra clans as *balbala*, gateways or doorways.

Bibliography

Abbink, J. 1993a. 'Ethnic Conflict in the "Tribal" Zone: the Dizi and Suri in Southern Ethiopia'. *Journal of Modern African Studies* vol. 31, no. 4: 675–82.

———— 1993b. 'Famine, Gold and Guns: the Suri of Southwestern Ethiopia, 1985–91'. *Disasters* vol. 17, no. 3: 218–25.

———— 1994. 'Changing Patterns of "Ethnic" Violence: Peasant – Pastoralist Confrontation in Southern Ethiopia and its Implications for a Theory of Violence'. *Sociologus* vol. 44, no. 1: 66–78.

———— 1998a. 'Violence and Political Discourse among the Chai Suri', in G.J. Dimmendaal and M. Last (eds), *Surmic Languages and Cultures*. Cologne: Köppe Verlag, 321–44.

———— 1998b. 'New Configurations of Ethiopian Ethnicity: the Challenge of the South'. *Northeast African Studies* vol. 5 (NS): 59–81.

———— 1999. 'Violence, Ritual and Reproduction: Culture and Context in Surma Dueling'. *Ethnology* vol. 38, no. 3: 227–42.

———— 2000a. 'Restoring the Balance: Violence and Culture among the Suri of Southern Ethiopia', in G. Aijmer and J. Abbink (eds), *Meanings of Violence: A Cross-Cultural Perspective*. Oxford and New York: Berg, 77–100.

———— 2000b. 'Tourism and Its Discontents. Suri-Tourist Encounters in Southern Ethiopia'. *Social Anthropology* vol. 8, no. 1: 1–17.

———— 2000c. 'Violence and the Crisis of Conciliation. Suri, Dizi and the State in South-west Ethiopia'. *Africa* vol. 70, no. 4: 527–50.

———— 2006. 'Discomfiture of Democracy? The 2005 Election Crisis in Ethiopia and Its Aftermath'. *African Affairs* vol. 105, no. 419: 173–200.

Agedew Redie and I. Hinrichsen (eds) 2002. *Self-Help Initiatives in Ethiopia*. Addis Ababa: GTZ OSHP Publication.

Agneta, F. and M. Tomassoli 1992. 'Pre-School Education and Priests' Schools in Pawe Resettlement Area', in P. Dieci and C. Viezzoli (eds), *Resettlement and Rural Development in Ethiopia: Social and Economic Research, Training and Technical Assistance in the Beles Valley*. Milan: Franco Angeli,

Aguilar, M. 1999. 'Pastoral Identities: Memories, Memorials, and Imaginations in the Postcoloniality of East Africa'. *Anthropos* vol. 94, no. 1–3: 149–61.

Alemayehu Haile, Boshi Gonfa, Daniel Deressa, Senbeto Busha and Umer Nurre 2004. *Oromo History Before the 16th Century: Oromiyaa Culture and Tourism Commisssion*. Addis Ababa: Artistic Printing Press.

Alexander Naty 1992. 'The Culture of Powerlessness and the Spirits of Rebellion Among the Ari People of Southwest Ethiopia'. PhD dissertation, Stanford University.

Amborn, H. 1976. 'Wandlungen im sozio-ökonomischen Gefüge der Bevölkerungsgruppen im Gardulla-Dobase-Horst in Südäthiopien'. *Paideuma* vol. 22: 151–61.

———— 1988. 'History of Events and Internal Development: the Example of the Burji-Konso Cluster', in Tadesse Beyene (ed.), *Proceedings of the VIIIth International Congress of Ethiopian Studies 1984*. Addis Ababa: Institute of Ethiopian Studies, 751–67.

———— 1989. 'Agricultural Intensification in the Burji-Konso Cluster of South-West Ethiopia'. *Azania* vol. 14: 71–83.

———— 1990. *Differenzierung und Integration, Vergleichende Untersuchungen zu Handwerkern und Spezialisten in südäthiopischen Agrargesellschaften*. Munich: Trickster.

———— 1994. 'Rethinking One's Own Culture (Emic and Etic Considerations)', in H.G. Marcus (ed.), *New Trends in Ethiopian Studies*. Lawrenceville, NJ: Red Sea Press, vol. 2., 773–90.

———— 1995. 'Von der Stadt zur sakralen Landschaft. Bóohée Burji. Eine städtische Siedlung in Südäthiopien. Von Helmut Straube, bearbeitet von Hermann Amborn. Helmut Straube zum 10.Todestag gewidmet'. *Tribus* vol. 44: 65–99.

———— 1997a. 'The Conceptionalization of Landscape in Southern Ethiopian Societies', in K. Fukui, E. Kurimoto and M. Shigeta (eds), *Ethiopia in Broader Perspective*. Kyoto: Shokado Booksellers, 379–88.

———— 1997b. 'Hunde Gottes: Eisenhandwerker und Demiurgen', in R. Klein-Ahrend, (ed.), *Traditionelles Eisenhandwerk in Afrika*. Colloquium Africanum 3. Cologne: Heinrich-Barth-Institut, 146–73.

———— 1998. 'Die zerfranste Ethnie. Zum analytischen Umgang mit komplexen Gesellschaften'. *Anthropos* vol. 93: 349–61.

———— 2006. 'The Contemporary Significance of What Has Been. Three Approaches to Remembering the Past: Lineage, Gada, and Oral Tradition'. *History in Africa* vol. 33: 53–84.

———— (forthcoming). *Burji in Äthiopien und Kenia*. Wiesbaden: Harrassowitz.

Amborn, H. and A. Kellner 1999. 'Burji Vocabulary of Cultural Items – An Insight into Burji Culture. Based on the Field Notes of Helmut Straube'. *Afrikanistische Arbeitspapiere* no. 58: 5–67.

Arendt, H. 1970. *On Violence*. New York: Harcourt, Brace and World.

Arero, H.W. 2007. 'Coming to Kenya: Imagining and Perceiving a Nation among the Borana of Kenya'. *Journal of Eastern African Studies* vol. 1, no. 2: 292–304.

Asmarom Legesse 1973. *Gada: Three Approaches to the Study of African Society*. New York: Free Press.

Bahrey 1954. *Some Records of Ethiopia: 1593–1646 / Being Extracts from the History of High Ethiopia or Abassia by Manuel de Almeida. Together with Bahrey's History of the Galla*. Trans. and ed. by C. F. Beckingham and G. W. B. Huntingford. London: Hakluyt Soc.

Bahru Zewde 1976. 'Relations between Ethiopia and the Sudan on the Western Ethiopian Frontier (1898–1935)'. PhD dissertation, University of London: School of Oriental and African Studies.

———— 1991. *A History of Modern Ethiopia 1855–1974*. Addis Ababa: Addis Ababa University Press.

Bartels, L. 1994. 'On Pilgrimage to a Holy Tree', in D. Brokensha (ed.), *A River of Blessings. Essays in Honor of Paul Baxter*. Syracuse, NY: Maxwell School of Citizenship and Public Affairs Syracuse University, 1–13.

Barth, F. 1969. *Ethnic Groups and Boundaries: the Social Organization of Culture Difference*. Bergen: Universitetsforlaget.

———— 1994. 'Enduring and Emerging Issues in the Analysis of Ethnicity', in H. Vermeulen and C. Govers (eds), *The Anthropology of Ethnicity: Beyond 'Ethnic Groups and Boundaries'*. Amsterdam: Het Spinhuis, 11–32.

———— 1998. 'Preface', in F. Barth, (ed.), *Ethnic Groups and Boundaries: the Social Organization of Culture Difference*. Long Grove, Ill.: Waveland Press, 5–8.

Bassi, M. 1999. 'The Complexity of a Pastoral African Polity: an Introduction to the Council Organisation of the Boran-Oromo'. *Journal of Ethiopian Studies* vol. 32: 15–33.

Baxter, P.T.W. 1978. 'Boran Age-sets and Generation-sets: *Gada* a Puzzle or a Maze?', in P.T.W. Baxter and U. Almagor (eds), *Age, Generation and Time: Some Features of East African Age Organization*. London: C. Hurst, 151–82.

——— 1979. 'Boran Age-Sets and Warfare', in K. Fukui and D. Turton (eds), *Warfare among East African Herders*. Senri Ethnological Studies. Osaka: National Museum of Ethnology, 69–94.

——— 1991. 'Preface' in J. Van De Loo (ed.), *Guji Oromo Culture in Southern Ethiopia: Religious Capabilities in Rituals and Songs*. Berlin: Dietrich ReimerVerlag, 9–13.

——— 1993. 'The "New" East African Pastoralist: An Overview', in J. Markakis (ed.), *Conflict and Decline of Pastoralism in the Horn of Africa*. London: Macmillan Press Ltd., 143–62.

Beke, C.T. 1844. 'Abyssinia – Being a Continuation of Routes in That Country'. *Journal of the Royal Geographical Society* vol. 14: 1–76.

——— 1845. 'On the Languages and Dialects of Abyssinia and the Countries to the South'. *Proceedings of the Philological Society* vol. II: 89–107.

Ben-Amos, D. 1996. 'Kontext', in K. Ranke and R.W. Brednich (eds), *Enzyklopädie des Märchens. Handwörterbuch zur historischen und vergleichenden Erzählforschung* vol. 8. Berlin: de Gruyter, 217–37.

Bender, M.L. (ed.) 1976. *The Non-Semitic Languages of Ethiopia*. East Lansing: African Studies Center.

Berihun Mebrate 1996. 'Spontaneous Settlements and Inter-Ethnic Relations in Metekel, Northwestern Ethiopia'. MA thesis in Social Anthropology, Addis Ababa University.

Berman, B. and J. Lonsdale 1992. *Unhappy Valley: Conflict in Kenya and Africa*. Book 1: *State and Class*. Athens, Ohio: Ohio University Press.

Black, P. and S. Otto 1973. *First Draft of a Konso Dictionary*. Murray Hill, NJ: Reproduced MS.

Blench, R. 2000. 'Nilo-Saharan Language Listing: Circulated for Comment'. http://homepage.ntlworld.com/roger_blench/Language%20data/NS%20language%20list.pdf, accessed 1 May 2008.

Blunt, P. and D.M. Warren, 1996. *Indigenous Organizations and Development*. London: IT Publications.

Bourdieu, P. 1983. 'Ökonomisches Kapital, kulturelles Kapital, soziales Kapital', in R. Kreckel (ed.), *Soziale Ungleichheiten*. Soziale Welt. Sonderband 2. Göttingen: Otto Schwarz, 183–98.

——— 1993. *Sozialer Sinn: Kritik der theoretischen Vernunft*. Frankfurt a.M.: Suhrkamp.

Bulatovich, A.K. 2000. *Ethiopia Through Russian Eyes: Country in Transition, 1896–1898*, translated and edited by R. Seltzer. Lawrenceville: Red Sea Press.

Casey, E. 1996. 'How to Get from Space to Place', in S. Feld and K. Basso (eds), *Senses of Place*. Sante Fe: School of American Research Press, 13–52.

Central Statistical Authority (CSA) 1996. *The 1994 Population and Housing Census of Ethiopia, Result for Southern Nations, Nationalities and Peoples' Region*, vol. 1, Part 1, *Statistical Report on Population Size and Characteristics*. Addis Ababa: Federal Democratic Republic of Ethiopia, Central Statistical Authority.

Central Statistical Authority (CSA)/Population and Housing Census Commission (PHCC) 1996. *The 1994 Population and Housing Census of Ethiopia: Results for Benishangul-Gumuz Region*. Volume I Statistical Report. Addis Ababa: CSA/PHCC.

Cerulli, E. 1933. *Etiopia Occidentale*, vol. II. Rome: Sindacato Italiano Arte Grafiche.

Chabal, P. 2001. 'Managing Disorder in Contemporary Africa: For a Realist Approach', in E. Kurimoto (ed.), *Rewriting Africa: Toward Renaissance or Collapse?* International Area Studies Conference VI, JCAS Symposium Series 14. Osaka: Japan Center for Area Studies, 11–25.

Chapple, D. 1998. 'The Firearms of Adwa', in Abdussamad H. Ahmad and R. Pankhurst (eds), *Adwa: Victory Centenary Conference 26 February–2 March 1996*. Addis Ababa: Addis Ababa University, Institute of Ethiopian Studies, 47–78.

Chisholm, J.S. 1993. 'Death, Hope and Sex. Life-history Theory and the Development of Reproductive Strategies'. *Current Anthropology* vol. 34, no. 1: 1–24.

Clapham, C. 1988. *Transformation and Continuity in Revolutionary Ethiopia*. Cambridge: Cambridge University Press.

——— 2002. 'Controlling Space in Ethiopia', in W. James, D.L. Donham, E. Kurimoto and A. Triulzi (eds), *Remapping Ethiopia: Socialism and After*. Oxford: James Currey, 9–30.

——— 2005. 'Comments on the Ethiopian Crisis'. http://www.ethiomedia.com/fastpress/clapham_on_ethiopian_crisis.html, accessed April 2006.

Cohen, A.P. 1994. 'Boundaries of Consciousness, Consciousness of Boundaries', in H. Vermeulen and C. Govers (eds), *The Anthropology of Ethnicity: Beyond 'Ethnic Groups and Boundaries'*. Amsterdam: Het Spinhuis, 59–79.

Cohen, R. 1978. 'Ethnicity: Problem and Focus in Anthropology'. *Annual Review of Anthropology* vol. 7: 379–403.

Cornwall, A. and N. Lindisfarne (eds) 1994. *Dislocating Masculinity: Comparative Ethnographies*. London: Routledge.

Darnell, A. and S. Parikh 1988. 'Religion, Ethnicity and the Role of the State: Explaining Conflict in Assam'. *Ethnic and Racial Studies* vol. 11, no. 3: 263–80.

Deleuze, G. and F. Guattari 1987. *A Thousand Plateaus: Capitalism and Schizophrenia*, translated by B. Massumi. Minneapolis: University of Minnesota Press.

Dessalegn Rahmato 1984. *Agrarian Reform in Ethiopia*. Uppsala: Scandinavian Institute of African Studies.

de Waal, A. 1997. *Famine Crimes: Politics and the Disaster Relief Industry in Africa*. Oxford: James Currey.

Dieci, P. and C. Viezzoli (eds) 1992. *Resettlement and Rural Development in Ethiopia: Social and Economic Research, Training and Technical Assistance in the Beles Valley*. Milan: Franco Angeli.

Donham, D.L. 1985. *Work and Power in Maale, Ethiopia*. Ann Arbor, Michigan: UMI Research Press.

——— 1986. 'Old Abyssinia and the New Ethiopian Empire: Themes in Social History', in D.L. Donham and W. James (eds.), *The Southern Marches of Imperial Ethiopia: Essays in History and Social Anthropology*. Cambridge: Cambridge University Press, 3–48.

——— 1990. *Work, Power, Ideology: Central Issues in Marxism and Anthropology*. Cambridge: Cambridge University Press.

——— 1994 *Work and Power in Maale, Ethiopia*, 2nd edn. New York: Columbia University Press.

——— 1999. *Marxist Modern: An Ethnographic History of the Ethiopian Revolution*. Berkeley: University of California Press; Oxford: James Currey.

——— 2002. 'Introduction', in D.L. Donham, W. James, E. Kurimoto and A. Triulzi (eds), *Remapping Ethiopia: Socialism and After*. Oxford: James Currey, 1–7.

Donham, D. L. and W. James (eds) 1986. *The Southern Marches of Imperial Ethiopia: Essays in History and Social Anthropology*. Cambridge: Cambridge University Press.

Duba Gololcha 1986. 'Warfare and Hunting among the Guji Oromo'. BA thesis, Department of Sociology and Social Administration. Addis Ababa University.

ECA (Economic Commission for Africa) 1998. *Benishangul Gumuz National Regional State SAERP-WARDIS: Forestry Report*. Addis Ababa: ECA.

Elwert, G. 1989. 'Nationalismus und Ethnizität: Über die Bildung von Wir-Gruppen'. *Kölner Zeitschrift für Soziologie und Sozialpsychologie* no. 3: 440–64.

——— 2002. 'Switching Identity Discourses: Primordial Emotions and the Social Construction of We-groups' in G. Schlee (ed.), *Imagined Differences: Hatred and the Construction of Identity*. Münster: LIT Verlag, 33–54.

Elwert, G., S. Feuchtwang, and D. Neubert (eds) 1999. *Dynamics of Violence: Processes of Escalation and De-Escalation in Violent Group Conflicts*. Berlin: Duncker and Humblot. *Sociologus* Beiheft 1.

Ethiopian Mapping Authority 1988. *National Atlas of Ethiopia*. Addis Ababa: Berhanena Selam Printers.

Evans-Pritchard, E. 1940. *The Nuer: A Description of the Modes of Livelihood and Political Institutions of a Nilotic People*. Oxford: Oxford University Press.

Evers, H.-D. and T. Schiel 1988. *Strategische Gruppen. Vergleichende Studien zu Staat, Bürokratie und Klassenbildung in der Dritten Welt*. Berlin: Dietrich Reimer Verlag.

Fabian, J. 1983. *Time and the Other: How Anthropology Makes its Object*. New York: Columbia University Press.

———— 1999. 'Remembering the Other: Knowledge and Recognition in the Exploration of Central Africa'. *Critical Inquiry* vol. 26, no. 1: 49–69.

Federal Democratic Republic of Ethiopia 1995. *Constitution*. Addis Ababa: Federal Negarit.

Finnegan, R. 1992. *Oral Traditions and the Verbal Arts. A Guide to Research Practices*. London, New York: Routledge.

Forster, J. 1969. 'Economy of the Gamu Highlands'. *Geographical Magazine* vol. 65: 429–37.

Freeman, D. and A. Pankhurst (eds) 2001. *Living on the Edge. Marginalised Minorities of Craftworkers and Hunters in Southern Ethiopia*. Addis Ababa: Addis Ababa University.

Fukui, K. 1979. 'Cattle Colour Symbolism and Intertribal Homicide among the Bodi' in K. Fukui, and D. Turton (eds), *Warfare among East African Herders*. Osaka: National Museum of Ethnology, 147–77.

———— 1984. 'Intertribal Relations through Conflict: The Pastoral Meken (Bodi) in South-West Ethiopia'. *Minzokugaku-Kenkyu* (The Japanese Journal of Ethnology) vol. 48, no. 4: 471–80.

———— 1988. 'Cultural Ideology and Ethnic Formation: Based upon Cases of Conflict in Bodi Society' in Junzo, K. and Fukui. K. (eds.), *What is the Ethnic Group?* Tokyo: Iwanami Shoten, 187–212 (in Japanese).

Fukui, K. and J. Markakis 1994. 'Introduction', in K. Fukui and J. Markakis (eds), *Ethnicity and Conflict in the Horn of Africa*. London: James Currey, 1–14.

Galaty, J.G. 1993. '"The Eye that Wants a Person, Where Can It Not See?": Inclusion, Exclusion and Boundary Shifters in Maasai Identity', in T. Spear, and R. Waller (eds), *Being Maasai*. London: James Currey, 174–94.

Garretson, P.P. 1986. 'Vicious Cycles: Ivory, Slaves, and Arms on the New Maji Frontier', in D.L. Donham and W. James (eds), *The Southern Marches of Ethiopia. Essays in History and Social Anthropology*. Cambridge: Cambridge University Press, 196–218.

Gebru Tareke 1991. *Ethiopia, Power and Protest: Peasant Revolts in the Twentieth Century*. Cambridge: Cambridge University Press.

Getachew, K.N. 1996. 'The Displacement and Return of Pastoralists in Southern Ethiopia. A Case Study of the Garri', in T. Allen (ed.), *In Search of Cool Ground: War, Flight and Homecoming in Northeast Africa*. London, Oxford, Trenton (NJ): Africa World Press, 111–23.

Gilmore, D.D. 1990. *Manhood in the Making: Cultural Concepts of Masculinity*. New Haven, London: Yale University Press.

Gluckman, M. 1966. *Custom and Conflict in Africa*. Oxford: Basil Blackwell.

Goffman, E. 1963. *Stigma: Notes on the Management of Spoiled Identity*. Englewood Cliffs, NJ: Prentice Hall.

Gray, S.J. 2000. 'A Memory of Loss: Ecological Politics, Local History, and the Evolution of Karimojong Violence'. *Human Organization* vol. 59, no. 4: 401–18.

Gufu Oba 1996. 'Shifting Identities along Resource Borders', in P.T.W. Baxter, J. Hultin and
    A. Triulzi (eds), *Being and Becoming Oromo*. Lawrenceville, NJ: Red Sea Press, 117–31.
Haberland, E. 1963. *Galla Süd-Äthiopiens: Ergebnisse der Frobenius-Expeditionen 1950–52 und
    1954–56*. Stuttgart: W. Kohlhammer Verlag.
Hall, S. 1996. 'Introduction: Who Needs "Identity"?', in S. Hall and P. du Gay (eds),
    Questions of Cultural Identity. London: Sage, 1–17.
Hallpike, C.R. 1972. *The Konso of Ethiopia: A Study of the Values of a Cushitic People*. Oxford:
    Clarendon Press.
Hann, C.M. 1996. 'Ethnic Cleansing in Eastern Europe: Poles and Ukrainians along the
    Curzon Line'. *Nations and Nationalism* vol. 2, no. 3: 389–406.
——— 1998. 'Postsocialist Nationalism: Rediscovering the Past in South East Poland'. *Slavic
    Review* vol. 57, no. 4: 840–63.
Haugerud, A. 1995. *The Culture of Politics in Modern Kenya*. Cambridge: Cambridge
    University Press.
Herskovits, M.J. 1926. 'The Cattle Complex in East Africa'. *American Anthropologist* vol. 28,
    no. 1: 230–72.
Hinnant T.J. 1977. 'The Gada System of the Guji of Southern Ethiopia'. PhD dissertation,
    University of Chicago.
Hodson, A.W. 1927. *Seven Years in Southern Abyssinia*. London: T. Fisher Unwin.
Homer-Dixon, T.F. 1999. *The Environment, Scarcity, and Violence*. Princeton: Princeton
    University Press.
Hutchinson, S.E. 1996. *Nuer Dilemmas. Coping with Money, War, and the State*. Berkeley, Los
    Angeles, London: University of California Press.
——— 2000. 'Nuer Ethnicity Militarized'. *Anthropology Today* vol. 16, no. 3: 6–13.
——— 2001. 'A Curse from God? Religious and Political Dimensions of the Post-1991 Rise
    of Ethnic Violence in South Sudan'. *Journal of Modern African Studies* vol. 39, no. 2:
    307–31.
Jackson, R.T. 1970. *Land Use and Settlement in Gamu Gofa, Ethiopia*. Occasional Paper No.
    17, Department of Geography. Kampala: Makerere University.
Jackson, R.T., P. Mulvaney, T.P.J. Russell and J.A. Forster 1969. 'Report of the Oxford
    University Expedition to the Gamu Highlands of Southern Ethiopia'. Unpublished
    report.
James, W. 1986. 'Lifelines: Exchange Marriage among the Gumuz', in D.L. Donham and W.
    James (eds), *The Southern Marches of Imperial Ethiopia*. Cambridge: Cambridge
    University Press. 119–47.
James, W., D.L. Donham, E. Kurimoto and A. Triulzi (eds) 2002. *Remapping Ethiopia:
    Socialism and After*. Oxford: James Currey.
Jenkins, R. 1996. 'Ethnicity Etcetera: Social Anthropological Points of View'. *Ethnic and
    Racial Studies* vol. 19, no. 4: 807–822.
Jensen, A.E. 1936. *Im Lande des Gada. Wanderungen zwischen Volkstrümmern Süd-Abessiniens*.
    Stuttgart: Strecker and Schröder.
——— (n.d.). 'Die Konso'. Manuscript.
Johnson, M. 1987. *The Body in the Mind*. Chicago: University of Chicago Press.
Joireman, S.F. 1997. 'Opposition Politics and Ethnicity in Ethiopia: We Will All Go Down
    Together'. *Journal of Modern African Studies* vol. 35, no. 3: 387–407.
Kassam, A. and Gemetchu Megerssa 1994. '*Aloof Alolla*: the Inside and the Outside: Boran
    Oromo Environmental Law and Methods of Conservation', in D. Brokensha (ed.), *A
    River of Blessings: Essays in Honour of Paul Baxter*. Syracuse, NY: Maxwell School of
    Citizenship and Public Affairs Syracuse University, 85–98.

Kellner, A. 2001. 'The Lazy Baboon: a Linguistic Analysis and Anthropological Interpretation of a Burji Tale'. *Afrikanistische Arbeitspapiere* no. 65: 41–69.

——— 2007. *Mit den Mythen denken. Die Mythen der Burji als Ausdrucksform ihres Habitus.* (With an English Summary). Hamburg: LIT.

Keupp, H. 1999. *Identitätskonstruktionen. Das Patchwork der Identitäten in der Spätmoderne.* Reinbek: Rowohlt.

Knutsson, K.E. 1967. *Authority and Change: A Study of the Kallu Institution among the Macha Galla of Ethiopia.* Etnologiska Studier 29. Göteborg: Ethnografiska Museet.

Kopytoff, I. 1986. 'The Cultural Biography of Things: Commoditization as Process', in A. Appadurai (ed.), *The Social Life of Things: Commodities in Cultural Perspective.* Cambridge: Cambridge University Press, 64–94.

Krylow, A. 1994. 'Ethnic Factors in Post-Mengistu Ethiopia', in Abebe Zegeye and S. Pausewang (eds), *Ethiopia in Change: Peasantry, Nationalism and Democracy.* London: British Academic Press, 231–41.

Kunstadter, P. 1978. 'Ethnic Group, Category and Identity: Karen in North-western Thailand', in C.F. Keyes (ed.), *Ethnic Adaptation and Identity: The Karen on the Thai Frontier with Burma.* Philadelphia: ISHI, 119–164.

Kuper, A. 1999. *Culture: The Anthropologists' View.* Cambridge, MA, London: Harvard University Press.

Kurimoto, E. 1992. 'Natives and Outsiders: the Historical Experience of the Anywaa of Western Ethiopia'. *Journal of Asian and African Studies* no.43: 1–43.

Kurimoto E. and S. Simonse (eds) 1998. *Conflict, Age and Power in North East Africa.* Oxford: James Currey.

Lakoff, G. and M. Johnson 1980. *Metaphors We Live By.* Chicago: University of Chicago Press.

Larebo, H.M. 1994. *The Building of an Empire: Italian Colonial Policy and Practice in Ethiopia 1935-1941.* Oxford: Clarendon Press.

Leach, E. 1954. *Political Systems of Highland Burma.* London: Athlone Press.

Leach, M. and R. Mearns (eds) 1996. *The Lie of the Land: Challenging Received Wisdom on the African Environment.* Oxford: James Currey.

Levinson, S. 1998. 'Studying Spatial Conceptualization across Cultures: Anthropology and Cognitive Science'. *Ethos* vol. 26, no. 1: 7–24.

Lewis, H.S. 1966. 'The Origins of the Galla and Somali'. *Journal of African History* vol. 7, no. 1: 27–46.

Lydall, J. 1994 'Beating around the Bush', in Bahru Zewde, R. Pankhurst and Taddese Beyene (eds), *Proceedings of the Eleventh International Conference of Ethiopian Studies, Addis Ababa, 1–6 April 1991.* Addis Ababa: Institute of Ethiopian Studies, Addis Ababa University, 205–25.

Lydall, J. and I. Strecker 1979a. *The Hamar of Southern Ethiopia,* vol. 1: *Work Journal.* Hohenschaftlarn: Klaus Renner Verlag.

——— 1979b. *The Hamar of Southern Ethiopia,* vol. 2: *Baldambe Explains.* Hohenschaftlarn: Klaus Renner Verlag.

Mamdani, M. 1996. *Citizen and Subject: Contemporary Africa and the Legacy of Late Colonialism.* London: James Currey.

Margalit, A. 1998. *The Decent Society.* Cambridge, MA: Harvard University Press.

Markakis, J. 1974. *Ethiopia: Anatomy of a Traditional Polity.* Oxford: Clarendon Press.

——— 1998. *Resource Conflict in the Horn of Africa.* Oslo: PRI; London: Sage Publications.

Masuda, K. 1997. 'Histories and Ethnic Identities among the Banna: an Analysis of the *Bitas'* Oral Histories', in K. Fukui, E. Kurimoto and M. Shigeta (eds), *Ethiopia in Broader*

Perspective: Papers of the 13th International Conference of Ethiopian Studies, vol. 2. Kyoto: Shokado Book Sellers, 456–70.

Matsuda, H. 1994. 'Annexation and Assimilation: Koegu and their Neighbours', in F. Katsuyoshi and J. Markakis (eds), *Ethnicity and Conflict in the Horn of Africa*. London: James Currey; Athens: Ohio University Press, 48–62.

——— 1997. 'How Guns Change the Muguji: Ethnic Identity and Armament in a Periphery', in K. Fukui, E. Kurimoto and M. Shigeta (eds), *Ethiopia in Broader Perspective: Papers of the 13th International Conference of Ethiopian Studies*, vol. 2. Kyoto: Shokado Book Sellers, 471–78.

Melesse Getu 1995. 'Tsemako Women's Roles and Status: In Agro-Pastoral Production'. PhD thesis, Addis Ababa University.

Merid Wolde Aregay 1971. 'Southern Ethiopia and the Christian Kingdom 1508–1708, with Special Reference to the Galla Migrations and their Consequences'. PhD thesis, University of London.

Miller, W.I. 1993. *Humiliation, and Other Essays on Honor, Social Discomfort and Violence*. Ithaca, London: Cornell University Press.

Mirzeler, M. and C. Young 2000. 'Pastoral Politics in the Northeast Periphery in Uganda: AK-47 as Change Agent'. *Journal of Modern African Studies* vol. 38, no. 3: 407–30.

Mude, K.A. 1969. 'The Amaro-Burji of Southern Ethiopia', in B. G. McIntosh (ed), *Ngano*. Nairobi Historical Studies 1. Nairobi: University of Nairobi, 26–47.

Metekel Zonal Agricultural Department (MZAD) 1992. 'Metekel, Chagni'. Pawe: MZAD (in Amharic).

——— 1999. 'Study on Basic Agricultural Data in Metekel Zone'. Pawe: MZAD (in Amharic).

Olmstead, J. 1997. *Woman Between Two Worlds: Portrait of an Ethiopian Rural Leader*. Urbana: University of Illinois Press.

Ottaway, M. and D. Ottaway 1978. *Ethiopia: Empire in Revolution*. New York: Africana Publishing Company.

Pankhurst, A. 1999. 'Castes in Africa: The Evidence from South-Western Ethiopia Reconsidered'. *Africa* vol. 69, no. 4: 485–509.

Pankhurst, A. and F. Piguet (eds) 2004. *People, Space and the State: Migration, Resettlement and Displacement in Ethiopia*. Addis Ababa: Ethiopian Society of Sociologists, Social Workers and Anthropologists.

Pankhurst, R. 1962. 'Fire-Arms in Ethiopian History (1800–1935)'. *Ethiopian Observer* vol. 6, no. 2: 135–80.

——— 1968. *Economic History of Ethiopia*. Addis Ababa: Haile Selassie I University Press.

——— 1985. *History of Ethiopian Towns from the Mid-nineteenth Century to 1935*. Wiesbaden: Franz Steiner Verlag.

——— 1990. *A Social History of Ethiopia*. Addis Ababa: Institute of Ethiopian Studies, Addis Ababa University.

——— 1997. *The Ethiopian Borderlands: Essays in Regional History from Ancient Times to the End of the 18th Century*. Lawrenceville, NJ: Red Sea Press.

Parker, G. 1988. *The Military Revolution: Military Innovation and the Rise of the West, 1500–1800*. Cambridge: Cambridge University Press.

Pausewang, S. 1994. 'Local Democracy and Central Control', in Abebe Zegeye and S. Pausewang, (eds), *Ethiopia in Change: Peasantry, Nationalism and Democracy*. London: British Academic Press, 209–30.

Popp, W.M. 2001. 'Yem. Janjero oder Oromo? Die Konstruktion ethnischer Identität im sozialen Wandel', in A. Horstmann and G. Schlee (eds), *Integration durch Verschiedenheit*. Bielefeld: transcript, 367–403.

Ranger, T. 1983. 'The Invention of Tradition in Colonial Africa', in E. Hobsbawm and T. Ranger (eds), *The Invention of Tradition*. Cambridge: Cambridge University Press, 211–62.

Sahlins, M.D. 1961. 'The Segmentary Lineage: an Organization of Predatory Expansion'. *American Anthropologist* vol. 63, no. 2: 322–45.

Sasse, H.-J. and H. Straube 1977. 'Kultur und Sprache der Burji in Süd-Äthiopien: Ein Abriß', in W.J.G Möhlig, F. Rottland and B. Heine (eds), *Zur Sprachgeschichte und Ethnohistorie in Afrika*. Berlin: Dietrich Reimer Verlag, 239–66.

Sato, S. 1996. 'The Commercial Herding System among the Garri', in S. Sato (ed.), *Essays in Northeast African Studies*. Senri Ethnological Studies no. 43. Osaka: National Museum of Ethnology,

Schlee, G. 1979. *Das Glaubens-und Sozialsystem der Rendille, Kamelnomaden Nordkenias*. Berlin: Dietrich Reimer Verlag.

—— 1988. 'Camel Management Strategies and Attitudes towards Camels in the Horn', in J.C. Stone (ed.), *The Exploitation of Animals in Africa*. Aberdeen: Aberdeen University African Studies Group, 143–54.

—— 1990a. *Policies and Boundaries: Perceptions of Space and Control of Markets in a Mobile Livestock Economy*. Working Paper No. 133. Bielefeld: University of Bielefeld, Sociology of Development Research Centre.

—— 1990b. 'Holy Grounds' in P.T.W. Baxter and R. Hogg (eds), *Property, Poverty and People: Changing Rights in Property and Problems of Pastoral Development*. Manchester: Department of Social Anthropology, University of Manchester, 45–54.

—— 1992. 'Ritual Topography and Ecological Use: The Gabbra of the Kenyan/Ethiopian Borderlands', in D. Parkin and E. Croll (eds), *Bush Base: Forest Farm*. London, New York: Routledge, 110–28.

—— 1994a [1989]. *Identities on the Move: Clanship and Pastoralism in Northern Kenya*. New York: St Martins Press; Nairobi: Gideon S. Were Press; Hamburg: LIT.

—— 1994b. 'Ethnicity Emblems, Diacritical Features, Identity Markers: Some East African Examples', in D. Brokensha (ed.), *A River of Blessings*. Syracuse, NY: Maxwell School of Citizenship and Public Affairs, Syracuse University. 129–43.

—— 1998a. 'Some Effects on a District Boundary', in M. Aguilar (ed.), *The Politics of Age and Gerontocracy in Africa*. Trenton, NJ, Asmara, Eritrea: Africa World Press, 225–56.

—— 1998b. 'Gada Systems on the Meta-Ethnic Level: Gabbra/Boran/Garre Interactions in the Kenyan/Ethiopian Borderland', in E. Kurimoto and S. Simonse (eds), *Conflict, Age and Power in North East Africa*. Oxford: James Currey, 121–46.

—— 1999. 'Nomades et l'état au nord du Kenya', in A. Bourgeot (ed.), *Horizons nomades en Afrique sahélienne*. Paris: Karthala, 219–39.

—— 2000. 'Identitätskonstruktionen und Parteinahme: Überlegungen zur Konflikttheorie'. *Sociologus* no. 1: 64–89.

—— 2002. 'Regularity in Chaos: the Politics of Difference in the Recent History of Somalia', in G. Schlee (ed.), *Imagined Difference: Hatred and the Construction of Identity*. Hamburg: LIT, 251–80.

—— 2007. 'Brothers of the Boran Once Again: On the Fading Popularity of Certain Somali Identities in Northern Kenya'. *Journal of Eastern African Studies* vol. 1, no. 3: 417–35.

Schlee, G. (in preparation). *Heilige Berge und Pilgerfahrten in Südäthiopien und Nordkenia*.

Schlee, G. and A.A. Shongolo 1995. 'Local War and its Impact on Ethnic and Religious Identifications in Southern Ethiopia'. *GeoJournal* vol. 36, no. 1: 7–17.

Schlee, G. and A.A. Shongolo 1996. 'Oromo Nationalist Poetry: Jarso Waago Qooto's Tape Recording About Political Events in Southern Oromia, 1991', in R.J. Hayward and I.M.

Lewis (eds), *Voice and Power: The Culture of Language in North-East Africa. Essays in Honor of B.W. Andrzejewski.* London: University of London, School of Oriental and African Studies, 229–42.

Schlee, G. and A.A. Shongolo (in preparation). *Islam and Ethnicity in Northern Kenya and Southern Ethiopia.*

Shongolo, A.A. 1996. 'The Poetics of Nationalism: A Poem by Jarso Waago Qoot'o', in P.T.W. Baxter, J. Hultin and A. Triulzi (eds), *Being and Becoming Oromo.* Uppsala: Nordiska Afrika Institut, 265–90.

Schott, R. 1968. 'Das Geschichtsbewußtsein schriftloser Völker'. *Archiv für Begriffsgeschichte* vol. 12: 166–205.

Simonse, S. 1992. *Kings of Disaster: Dualism, Centralism and the Scapegoat King in Southeastern Sudan.* Leiden, New York, Copenhagen, Cologne: E.J. Brill.

Smith, G. (ed.) 1996. *Federalism: the Multi-Ethnic Challenge.* London: Longman.

Sobania, N. 1979. *A Background History to the Mount Kulal Region of Northern Kenya.* Nairobi: UNESCO.

––––– 1980. 'The Historical Tradition of the Peoples of the Eastern Lake Turkana Basin, c. 1840–1925'. PhD dissertation, SOAS, University of London.

––––– 1988a. 'Pastoral Migration and Colonial Policy' in D. Anderson and D. Johnson (eds), *The Ecology of Survival: Case Studies from Northeast African History.* London: Crook Acad.Publ., 219–40.

––––– 1988b. 'Fishermen Herders: Subsistence, Survival and Cultural Change in Northern Kenya'. *Journal of African History* vol. 29, no. 1: 41–56.

Spear, T. 1993. 'Introduction', in T. Spear and R. Waller (eds), *Being Maasai: Ethnicity and Identity in East Africa.* Oxford: James Currey, 1–18.

Spear, T. and R. Waller (eds) 1993. *Being Maasai: Ethnicity and Identity in East Africa.* London: James Currey.

Spencer, P. 1973. *Nomads in Alliance: Symbiosis and Growth among the Rendille and Samburu of Kenya.* Oxford: Oxford University Press.

Straube, H. n.d., c. 1955. 'Die Burdji'. Manuscript.

––––– 1963. *Westkuschitische Völker Süd-Äthiopiens.* Stuttgart: W. Kohlhammer Verlag.

Tablino, P. 1999. *The Gabra. Camel Nomads of Northern Kenya.* Marsabit: Paulines. (Corrected, supplemented and reorganized edition of 'I Gabbra del Kenya', 1980 Bologna: EMI).

Taddesse Berisso 1988. 'Traditional Warfare among the Guji of Southern Ethiopia'. MA thesis (Anthropology), Michigan State University, East Lansing.

––––– 1995. 'Agricultural and Rural Development Policies in Ethiopia: a Case Study of Villagization Policy among the Guji-Oromo of Jam Jam Awraja'. PhD dissertation, Michigan State University.

Taddese Tamrat 1988. 'Nilo-Sahara Interactions with Neighbouring Highlanders: The Case of the Gumuz of Gojjam and Wollega', in *Proceedings of the Workshop on Famine Experience and Resettlement in Ethiopia. 29–30 December 1988.* Addis Ababa: Institute of Development Research, Addis Ababa University, 7–15.

Tadesse Wolde Gossa 1991. 'Some Social and Ritual Functions of Gamo and Konso Public Places' in *Proceedings of the Eleventh International Conference of Ethiopian Studies.* Addis Ababa University Press, 325–40.

––––– 1999. 'Warfare and Fertility: a Study of the Hor (Arbore) of Southern Ethiopia'. PhD dissertation, London School of Economics and Political Sciences, Department of Anthropology.

————— 2002. 'Evading the Revolutionary State: the Hor under the Derg', in W. James, D.L. Donham, E. Kurimoto and A. Triulzi (eds), *Remapping Ethiopia: Socialism and After.* Oxford: James Currey, 37–58.

Tafla, B. (ed.) 1987. *Asma Giyorgis and His Work: History of the Galla and the Kingdom of Shoa.* Stuttgart: Franz Steiner Verlag.

Tedecha Gololcha 1988. 'The Politico-Legal System of the Guji Oromo'. BA thesis (Law), School of Law, Addis Ababa University.

Tornay, S. 1993. 'More Chances on the Fringe of State? The Growing Power of the Nyangatom, a Border People of the Lower Omo Valley, Ethiopia (1970–1992)', in T. Tvedt (ed.), *Conflicts in the Horn of Africa: Human and Ecological Consequences of Warfare.* Uppsala: Uppsala University, 143–63.

————— 1995. 'Structure et événement: le système générationnel des peuples du cercle Karimojong'. *L'Homme* vol. 134: 51–80.

————— 1997. 'Les visions de Natubwae, ou l'imaginaire sacrificiel d'une néophyte', in J. Hainard and R. Kaehr (eds), *Dire les autres, réflexions et pratiques ethnologiques.* Lausanne: Payot, Sciences Humaines, 309–24.

————— 2001. *Les fusils jaunes. Générations et politique en pays Nyangatom (Éthiopie).* Nanterre: Société d'Ethnologie.

Tornay, S. and E. Sohier 2007. *Empreintes du temps. Le sceaux des dignitaires éthiopiens du règne de Téwodros à la régence de Täfäri Mäkonnen.* Addis Ababa: Centre français d'études éthiopiennes et Institute of Ethiopian Studies.

Triulzi, A. 1981. *Salt, Gold and Legitimacy: Prelude to the History of a No-man's Land – Bela Shangul, Wallaga, Ethiopia.* Naples: Instituto Universitario Orientale, Seminario di Studi Africani.

Tronvoll, K. 2001. 'Voting, Violence and Violations: Peasant Voices on the Flawed Elections in Hadiya, Southern Ethiopia'. *Journal of Modern African Studies* vol. 39, no. 4: 697–716.

Turner, V. 1995. *The Ritual Process: Structure and Anti-structure.* Chicago: Aldine Press.

Turton, D. 1993. '"We Must Teach Them to be Peaceful": Mursi Views on Being Human and Being Mursi', in T. Tvedt (ed.), *Conflicts in the Horn of Africa: Human and Ecological Consequences of Warfare.* Uppsala: Uppsala University, 164–80.

————— 1997. 'War and Ethnicity: Global Connections and Local Violence in North East Africa and Former Yugoslavia'. *Oxford Development Studies* vol. 25, no. 1, 77–94.

van Gennep, A. 1960. *The Rites of Passage.* Chicago: University of Chicago Press.

Vansina, J. 1985. *Oral Tradition as History.* London: James Currey.

von Clausewitz, K. 1908. *On War.* London. Kegan Paul and Co.

Watson, E.E. 1998. 'Ground Truths: Land and Power in Konso'. PhD thesis, University of Cambridge.

————— 2002. 'Capturing a Local Elite: the Konso Honeymoon', in J. Wendy, D.L. Donham, E. Kurimoto and A. Triulzi (eds), *Remapping Ethiopia: Socialism and After.* Oxford: James Currey, 198–218.

————— 2004. 'Agricultural Intensification and Social Stratification: Konso contrasted with Marakwet', in M. Widgren and J.E.G. Sutton (eds), *Islands of Intensification in Eastern Africa.* Oxford: James Currey; Nairobi: British Institute in East Africa, 49–67.

————— 2006. 'Making a Living in the Post-socialist Periphery: Struggles Between Farmers and Traders in Konso, Ethiopia'. *Africa* vol. 76, no. 1: 70–87.

Watson, E.E. and Lakew Regassa 2001. 'Konso', in D. Freeman and A. Pankhurst (eds), *Living on the Edge: Marginalised Minorities of Craftworkers and Hunters in Southern Ethiopia.* Addis Ababa: Addis Ababa University, 246–64.

Weber, M. 1976 [1922]. *Wirtschaft und Gesellschaft. Grundriss der Verstehenden Soziologie 5*, rev. edn. Tübingen: Mohr.

Widgren, M. and J.E.G. Sutton (eds) 2004. *Islands of Intensive Agriculture in Eastern Africa*. Oxford: James Currey.

Wolde-Selassie Abbute 1996. 'Report on a Mission to Identify Facts about Rural Community Water Supply in Region Six (Assosa)'. Mimeo. Addis Ababa: CISP.

———— 2000. 'Social Re-articulation after Resettlement: Observing the Beles Valley Scheme in Ethiopia', in M.M. Cernea, and C. McDowell (eds), *Risks and Reconstruction: Experiences of Resettlers and Refugees*. Washington, DC: World Bank, 412–30.

Wood, John. 1999. *When Men are Women: Manhood Among Gabra Nomads of East Africa*. Madison: University of Wisconsin.

Young, J. 1996. 'Ethnicity and Power in Ethiopia'. *Review of African Political Economy* vol. 23, no. 70: 531–42.

———— 1999. 'Along Ethiopia's Western Frontier: Gambela and Benishangul in Transition'. *Journal of Modern African Studies* vol. 37, no. 2: 321–46.

Zerihun Mohammed 1999. 'Natural Resource Competition and Interethnic Relations in Wondo Genet'. MA thesis in Social Anthropology, Addis Ababa University.

Zitelmann, T. 1999. 'Des Teufels Lustgarten: Themen und Tabus der politischen Anthropologie Nordostafrikas'. Habilitation thesis, Freie Universität Berlin: Institut für Ethnologie.

Notes on Contributors

Jon Abbink is Senior Researcher at the African Studies Centre, Leiden, the Netherlands, and Professor of African Ethnic Studies at the VU University in Amsterdam. His research is on the anthropology and history of Ethiopia, the comparative study of ethnic relations in North-East Africa and on political culture and conflict in North-East Africa. His latest book (edited with A. van Dokkum) is *Dilemmas of Development. Conflicts of Interest and their Resolution in Modernizing Africa* (Leiden: ASC, 2008).

Hermann Amborn is Professor Emeritus in the Institut für Ethnologie, Universität München. He completed his PhD in 1973, and his habilitation in 1987. His habilitation thesis was published as *Differenzierung und Integration* (1990) and was a study of the division of work in polycephalous societies. He has carried out research in the Konso-Dullay-Burji cluster since 1973 in Ethiopia, and in Kenya since 1981. Shorter research periods have been spent in Pakistan and Indonesia. He has taught in universities in Munich, Hamburg, Berlin, Tübingen and Manhattan (Kansas). His publications relate to polycephalous societies, identity, conflict solving, action research and ethics.

Alexander Kellner took his MA in 1995, studying cultural anthropology, folklore/European ethnology and sociology at the Ludwigs-Maximilians-Universität in Munich. His PhD thesis on Burji oral literature was published as *Mit den Mythen denken. Die Mythen der Burji als Ausdrucksform ihres Habitus* (With an English Summary) (2007) and is a study of how Burji in Kenya and Ethiopia use their oral traditions as a reflexive medium. For this purpose he combined hermeneutic-interpretative methods with the praxeological approach of Pierre Bourdieu. At present he is a lecturer at the Institut für Ethnologie, Universität München. His main fields of research include oral traditions (and history), ethnolinguistics and polycephalous societies in Ethiopia and Kenya.

Ken Masuda is Associate Professor in the Department of Environmental Science, Nagasaki University, and received his PhD in social Anthropology from Tokyo Metropolitan University. His dissertation has been published as *Being Peripheral: On Narratives of Modernity in Southern Ethiopia*. He has been carrying out field research among the Banna of southern Ethiopia with a special focus on interethnic conflicts and modernization process.

Günther Schlee is a Director at the Max Planck Institute for Social Anthropology at Halle. Until 1999 he was a Professor at Bielefeld. His habilitation thesis (Bayreuth, 1986) has been published as *Identities on the Move: Clanship and Pastoralism in*

Northern Kenya in the series of the International African Institute (1989). In his department, Integration and Conflict, he directs research in Africa, Central Asia and Europe.

Taddesse Berisso is Associate Professor of Anthropology at the Institute of Ethiopian Studies, Addis Ababa University. He has published several articles on the Guji-Oromo.

Tadesse Wolde Gossa holds a BA in Ethiopian Languages and Literature from Addis Ababa University and an MSc and PhD in Anthropology from the London School of Economics, with a thesis entitled 'Warfare and Fertility: a Study of the Hor (Arbore) of Southern Ethiopia'. His research interests are pastoralism, warfare, indigenous religious traditions and economic networks. He has published and edited a number of articles and books including 'Evading the Revolutionary State' in *Remapping Ethiopia: Socialism and After* (eds W. James, D. Donham, Eisei Kurimoto and A. Triulzi, Oxford: James Currey, 2002); 'Time and Religious Practice in Southwest Ethiopia', in *The Qualities of Time: Anthropological Approaches* (eds W. James and D. Mills, Oxford: Berg, 2005); (with T. Widlok) *Property and Equality. Vol. I. Ritualisation, Sharing, Egalitarianism* (New York: Berghahn, 2005); (with T. Widlock) *Property and Equality. Vol. II. Encapsulation, Commercialisation, Discrimination* (New York: Berghahn, 2005). He has also been involved in the production of two films about Konso.

Serge Tornay is Professor Emeritus at the Muséum National d'Histoire Naturelle, Paris, France. He studied humanities in Fribourg (Switzerland) and Vienna (Austria) then prehistory, linguistics and social anthropology in Paris. He taught anthropology first at the Sorbonne (1969–71) and then at the Laboratoire d'ethnologie et de sociologie comparative in the University of Nanterre (1971–96). In 1996 he joined the French National Museum of Natural History as Professor of Anthropology. In 1998–99, he was Director of the Musée de l'Homme. As a fieldworker, he studied first linguistics among the Keyo of Kenya, with a thesis on Kalenjin language in 1970. Since 1970, he has conducted ethnographic fieldwork among the Nyangatom of the Lower Omo valley, South-western Ethiopia, and among related peoples of north-western Kenya and south-eastern Sudan. He described the generation system of the Nyangatom as the key to their political organization. He also edited a collection of essays on colour perception, naming and symbolism, and published many papers on various themes of political and religious anthropology. In 2001 he published *Les Fusils jaunes*, a synthesis of his ethnographic work among the Nyangatom. In cooperation with Estelle Sohier he published in Addis Ababa *Empreintes du temps. Les sceaux des dignitaries éthiopiens du règne de Téwodros à la régence de Täfäri Mäkonnen.* The work received the 2007 Award of the Académie des Sciences d'outre-mer, Paris.

Elizabeth E. Watson is a Lecturer in the Department of Geography at the University of Cambridge. Her first degree is in anthropology and her second in geography. Her research focuses on environment and development issues, mainly in Eastern Africa, and most of her work in Ethiopia has been among the Konso. Recent publications include: 'Local Community, Legitimacy, and Cultural Authenticity in Postconflict Natural Resource Management: Ethiopia and Mozambique', *Environment and Planning D: Society and Space*, 2006 (with R. Black), and 'Making a Living in the Post-Socialist Periphery: Struggles between Farmers and Traders in Konso, Ethiopia', *Africa*, 2006.

Wolde-Selassie Abbute obtained his PhD in Social Anthropology from the University of Göttingen in 2002. He has published a number of articles on ethnic relations, resettlement, rural livelihoods, food security, marginalized minorities, social institutions and indigenous knowledge. His most recent book publication is *Gumuz and Highland Resettlers: Differing Strategies of Livelihood and Ethnic Relations in Metekel, Northwestern Ethiopia* (Münster, Germany: LIT Publishers, 2004).

John C. Wood has worked with Gabra pastoralists in northern Kenya and southern Ethiopia since 1991. He teaches anthropology at the University of North Carolina at Asheville. His research interests include gender and ethnicity, memory and meaning. His book *When Men are Women: Manhood among Gabra Nomads of East Africa*, Madison: University of Wisconsin, was published in 1999. He recently completed work on a partial sequel to that book, called *The Space Between*, on Gabra death rituals.

Index